Vegetables
Proceedings of the Oxford Symposium on Food and Cookery 2008

Vegetables

Proceedings of the Oxford Symposium on Food and Cookery 2008

Edited by Susan R. Friedland

Prospect Books
2009

First published in Great Britain in 2009 by Prospect Books, Allaleigh House, Blackawton, Totnes, Devon, TQ9 7DL.

© 2009 as a collection Prospect Books.
© 2009 in individual articles rests with the authors.

The authors assert their moral right to be identified as authors in accordance with the Copyright, Designs & Patents Act 1988. No part of this publication may be reproduced, stored in a retrieval system or transmitted in any form of by any means, electronic, mechanical, photocopying, recording or otherwise, without the prior permission of the copyright holders.

ISBN 978-1-903018-66-8

The photograph on the front cover is of a market in Cuenca, Ecuador. © 2009 Clare Pawley. That on the back cover is a plate from the *Flora Danica* set from Den Kongelige Danske Porcelænsfabrik illustrating the paper by Anna Marie Fisker and Tenna Doktor Olsen, photograph Anna Marie Fisker.

Design and typesetting in Gill Sans and Adobe Garamond by Tom Jaine.
Printed and bound in Great Britain by The Cromwell Press Group, Trowbridge.

Contents

Vegetable Carving: For Your Eyes Only   9
Julia Abramson

The War of the Vegetables: The Rise and Fall of the English
Allotment Movement   19
Lesley Acton

The First Scientific Defense of a Vegetarian Diet   29
Ken Albala

The Roman Vegetable Garden   36
Joan P. Alcock

The Bitter – and Flatulent – Aphrodisiac: Synchrony and Diachrony
of the Culinary Use of Muscari Comosum in Greece and Italy   46
Anthony F. Buccini

We Talked About the Aubergines: International Diplomacy
and the Cretan Diet   56
Andrew Dalby

The Carrot Purple   63
Joel S. Denker

Listening to Vegetables   71
Len Fisher and Nick Sorensen

Vegetables as a Symbol in Design and Art   77
Anna Marie Fisker & Tenna Doktor Olsen

An Edible Wild Thistle from the Lebanese Mountains   83
Anissa Helou

Allotment Diaries   85
Phil Iddison

Salvation in Sweetness? Sugar Beets in Antebellum America   95
Cathy K. Kaufman

Up on the Farm: The Role of Vegetables in Conquering Space   105
Jane Levi

The History of the Potato in Irish Cuisine and Culture   111
Máirtín Mac Con Iomaire and Pádraic Óg Gallagher

'Sweet as'– Notes on the Kumara or New Zealand
Sweet Potato as a *Taonga* or Treasure 121
Ray McVinnie

The American Pumpkin 132
Mark McWilliams

Wild Thing: The Naga Morich Story 139
Michael Michaud and Joy Michaud

'Per rape et porri et per spinachi': Examining the Realities of
Vegetable Consumption at the Monastery of Santa Trinità
in Post-Plague Florence 146
Salvatore Musumeci

The *Maraîchers* – Market Gardeners of the Ile-de-France 156
Lizabeth Nicol

The Southern California Vegetable Cult 166
Charles Perry

From the Plate to the Palate: Visual Delights from the
Vegetable Kingdoms of Italy 171
Gillian Riley

The Still-life Painter 180
Alicia Ríos; translated by Raymond Sokolov; photographs by Johanna Hecht

But, Did the English Eat Their Vegetables? A Look at English Kitchen
Gardens, and the Vegetable Cookery they Imply (1650–1800) 184
William Rubel

Who Put the Leeks in Cock-a-leekie Soup? 194
Allyson E. Sgro

Bone-dry Freshness: Dried Vegetables 202
Aylin Öney Tan and Filiz Hösükoğlu

Dokonjo Daikon: The Radish with the Fighting Spirit 209
Michelle Toratani

The Pomtajer 216
Karin Vaneker

A Vegetable Zodiac from Late Antique Alexandria 225
Susan Weingarten

Vegetable Carving: For Your Eyes Only

Julia Abramson

In America and Europe, ornate vegetable carving is in fashion. But just what is ornate vegetable carving? Food carving more generally is a familiar topic, because we carve meat. In those cultures that structure meals around meat, meat carving forms part of the universal cultural grammar. By contrast, vegetable carving has been little practiced and infrequently discussed. Yet in the last three decades, and particularly in the last few years, it has seized the attention of cooks and eaters, practitioners and consumers, authors and readers, amateurs and professionals alike.*

Today's sophisticated food economy and general climate of affluence place a premium on culinary artistry. Vegetable carving responds to this requirement. But ornate vegetable carving also takes food craft in unexpected directions. Consider recent feats of creativity in cooking. The molecular enchantments of Basque chef Ferran Adrià and the playful edible puns of Napa Valley culinary 'monk' turned New York gastro-pope Thomas Keller hold our attention because they stimulate the intellect as well as the senses. To produce an exquisite, complex vegetable carving or vegetable sculpture is, similarly, to accomplish a feat of unusual artistry. But Adrià and Keller are cooks. A cook's creations feed hungry stomachs. Similarly, meat carving breaks down carcasses, assisting us to eat and even beginning the process of digestion. By contrast, we consume ornate vegetable carving with our eyes. The carved vegetable does not feed our stomachs. In fact, vegetable carving participates in a much larger trend to re-allocate food resources for purposes other than consumption as food. The trend for vegetable carving occupies a niche in the more rarefied realms of the food world, and yet, as we shall see, the practice is exemplary.

This essay establishes a chapter in the history of carving to account for the recent trend in vegetable sculpture, define its characteristics, and analyze its cultural meaning. The recent emergence of ornate vegetable carving underlines predominant beliefs, values, and desires that shape our society, while also pointing to fault lines in our ideologies.

******

Food, of course, *is* nourishment. Or is it? In vegetable sculpture, the usual relationship of food to nourishment gets turned on its head. Consider, in Yuci Tan's recent *The Art*

---

*   I thank the friends and colleagues who read and responded to early versions of this essay: Julia Ehrhardt, Priscilla Parkhurst Ferguson, Alison Matthews-David, Christina Parte, and Carolin C. Young. Ning Yu provided invaluable help by translating and discussing with me Chinese-language source materials.

Vegetable Carving: For Your Eyes Only

of Food Sculpture: Designs and Techniques (2002), the entwined couple carved from an ochre butternut squash. She clothed in a wide, long skirt, he in a formal suit, spiral up out of vegetable flesh, which gives them substance. ('An Embrace,' p. 130) Tan's squash sculpture beautifully underlines the exceptional marriage of medium and function in vegetable carving. The butternut squash, *Cucurbita moschata*, native to Mexico, and relatives such as the ancestral neck pumpkin, are suited for carving. A developed squash presents a hefty, solid volume – a pumpkin hunk of marble awaiting its Praxiteles. The firm flesh resists the knife, allowing the carver precision and range. The squash sweats when peeled and cut, but it is slow to discolor with oxidation. Carved squash retains its integrity longer than most cut vegetables or fruits before, inevitably, it rots or desiccates. As a food, the butternut squash usefully provides calories, fiber, A vitamins. Its soft texture and sweet taste please the palate. In the pot, squash amiably combines with a host of other flavors from onions and sage to apples to soy sauce, and it adapts easily to a range of cooking methods. Butternut squash is astringent as well as hard when raw. It must be eaten cooked. The carved wedding squash, like nearly all vegetable sculptures, is raw, although highly finished. Processing of foods through cooking makes them palatable. The same is not true for carving vegetables.

Vegetable carving shifts the site of consumption, the locus of appreciation, from the mouth to the eye. The wedding squash has – visibly – been extensively handled by the carver. We see it differently, not as food, but rather as a visual object. We may admire or critique the carver's artistry. We may worry that, heavily manipulated, the raw squash has perhaps begun to spoil, due also to the exposure that allows us to take it in visually. Vegetable carving, like cooking and meat carving, takes foods as its basic materials. Unlike both cooking and meat carving, it rejects the edible character of those foods. In effect, the carved squash is unavailable as food. It will not feed our hunger. We eat up vegetable sculpture with our eyes only. It transforms us into observers, connoisseurs of the visual. Unlike cooking and meat carving, vegetable sculpture privileges looking over tasting. It chooses art and representation over life.

Until quite recently, vegetable carving in the West was a weak tradition, relative to the strong practice of meat carving. Ornate vegetable carving can be traced back in Western print sources to the European Renaissance with its manuals on meat carving. In these texts, vegetable carving sounds a faint counterpoint against the dominant theme of meat carving. A notable milestone in both histories is the most influential carving treatise published in Continental Europe, the Bavarian Matthias Geigher's *Il Trinciante* [*The Carver*, 1621]. Like a handful of authors before him who wrote about carving, Giegher devotes most space in his manual to narrating how to carve and serve meat. He also takes the innovative step of including technical diagrams that visually show how to carve a joint of meat, a game bird, a fish. Copied and imitated in a plethora of other volumes, Giegher's diagrams provide the paradigm for every illustrated Western work on meat carving published since 1621. At the very end of his treatise, Giegher tacked on a few pages of striking illustrations for carving 'Frutti' or fruit. These figures show the imaginative, largely abstract transformations that can be performed on the round and ovoid geometries of an orange or a citron. A few figural carvings charm fruits into shellfish and four-legged animals. The detail on the double-headed eagle (the east- and west-facing symbol of the Habsburg Empire) would have been even easier to realize using a large, hard vegetable.

Giegher's illustrations clue us in to the different status accorded to meat versus to fruit and vegetable carving. Meat carving and serving – which occupy the bulk of his pages and of other carving manuals – were an essential part of a formal meal. The (meat) carver's gestures ostentatiously demonstrated the host's status, enforced hierarchies among guests at table, and – not least – delivered animal protein to each guest. Clearly, fruit and vegetable carving have a long history in the West. Yet the carving manuals show us that it has not enjoyed the same high status as meat carving and was never an essential feature of meal service.

Several elements underline the secondary role of fruit and vegetable carving. Illustrations for vegetable and fruit carving appear in only a few early carving texts, where they are relegated to the last pages. In Giegher's manual, and in the carving section of the French anthology *L'École parfaite des officiers de bouche* [*Finishing School for Table Officers*, 1662], the diagrams for carving non-meat items have no narrative instructions like those that habitually accompany figures for meat carving. Nor do the carving manuals articulate what role carved vegetables and fruits were to play in a meal. Were fruits and vegetables carved decoratively prior to meal service? Were they rather slashed tableside as another virtuoso *coup* to divert diners? The early modern manuals, fulsome on meat matters, provide no answer to these questions.

With their lack of concerted attention to fruit and vegetable carving, the European manuals imply that the practice was superfluous in the modernizing gastro-culinary climate. By contrast, meat carving instructions reached an even larger readership as cookbooks began simultaneously to multiply and to subsume discussions of carving. By the eighteenth century, meat carving instructions or manuals contained within cookbooks reached an ever-increasing, and increasingly bourgeois, audience. Fruit and vegetable carving largely disappeared, both from European carving manuals and from cookbooks having specific sections devoted to the art of carving. The near-total disappearance reflects changing values in early modern society. Eighteenth-century French cookbooks valued practicality if not frugality. The cookbooks aim at an audience of non-elite as well as elite readers. Aristocratic display, with its ostentation and redundancy, is no longer essential. Rather, values of comfort, quality, taste, practicality, and enjoyment gain the upper hand. Cook-carvers now focus intently on the needs of the stomach.

A practical climate in a meat-eating culture leaves little room for fantastical flourishes such as fruit and vegetable carving. The first and most widely influential carving treatise of the French post-Revolutionary era makes no mention whatsoever of vegetable and fruit carving. The omission from Grimod de la Reynière's *Manuel des amphitryons* [*Manual for Hosts*, 1808] was surely deliberate. An aim of his *Manuel* was to revive but also to modernize old dining custom following the Revolution of 1789. From personal experience and from the books in his library, the author commanded a deep knowledge of old-regime aristocratic dining custom. Grimod was as aware of the early modern tradition of vegetable and fruit carving as of the art of meat carving that he revised based on old models. His omission of vegetable carving from the *Manuel* is telling. The omission confirms that the early nineteenth-century non-aristocratic table had little space for merely decorative elements. The suppression of the superfluous, and a long historical slumber for ornate fruit and vegetable carving, were the price of a broader continuity in cultural practice.

******

In the last thirty years, the Western food economy has re-appraised, and in fact re-invented, ornate vegetable carving. In America but also in Europe, a new generation of specialized fruit and vegetable carving manuals has blossomed since the 1980s. The guides are numerous and varied. Simple, whimsical, precious, gorgeous, professional, exquisite – the new carving books exhibit the range of styles and tones. They showcase virtuosic qualities, creative imagination, playfulness, quirky humor. Combinations of material, subject, and context veer off into the realm of kitsch to produce a work that we can call beautiful. Gail Gibbons's homespun *Pumpkin Book* caters to children, the youngest Halloween revelers. The elegantly minimalist, warmly creative art of assemblage photographed in Joost Elffers's *Play with Your Food* is designed for children, but what adult could fail to be charmed? The majority of works present a more complex art of carving. *Edible Art*, by Narahenapitage Sumith Premalal De Costa, speaks to the creative home chef. The Rosen brothers' *Culinary Carving and Plate Decorating* illustrates highly intricate sculptures. Yuci Tan's *Art of Food Sculpture* encourages vegetable carving as an art therapy. The hefty, quadri-lingual *Grosse Lehrbuch der Gemüse- & Früchte-Schnitzerei* [*Complete Manual of Vegetable and Fruit Carving*], by Xiang Wang and Louise & Othmar Fassbind, speaks to professional chefs and caterers and to ambitious, accomplished amateurs. Omitting nothing, Wang, Fassbind, and Fassbind's *oeuvre* even includes a text on hand surgery (in case of injury while carving) written by a medical doctor. A special sensitivity to the natural attributes of the fruits and vegetables that serve as base materials, along with outstanding cutting technique, enhance the carvings in Nidda Hongwiwat's *Vegetable and Fruit Carving*. The carvings on Hongwiwat's pages present an unusually refined esthetic.

The new works on vegetable and fruit carving combine step-by-step technical instruction with illustrations of finished, assembled works. Illustrations and especially photographs of completed vegetable carvings are a key feature, certainly an attractive one, in the literature. Vegetable carving manuals differ in this sense from works on meat carving. Guides to meat carving rarely show all steps in the carving process, nor its result. The end of meat carving is a broken down roast, a fowl that is reduced to small pieces. By contrast, vegetable carving often adds, assembles, constructs, although it may subtract and divide for detail. We see and talk about a vegetable carving as finished, complete, integral.

Autonomy is a distinguishing feature of the new vegetable carving. It is true that the craft of the new vegetable carving twins that of the more widely familiar plate garnish. The new vegetable carvings are as ephemeral and perishable as their humbler garnish cousins. Both a small garnish and a fully developed vegetable carving creatively capitalize upon the intrinsic features of the raw materials. But the new vegetable sculpture differs in intention and function from the garnish, which it also surpasses in complexity. The garnish plays a secondary role to food that it accompanies. The new carving is an end in itself.

The autonomous vegetable carving or sculpture aligns itself with and fulfills the varied functions of the work of art. Like works of plastic art realized in any other media, the new vegetable sculpture exists both as an aesthetic object that we admire and as a commodity that commands a market price. Vegetable carving offers possibilities for aesthetic, intellectual, and emotional expression and for practical exploitation that are unavailable to the humdrum garnish. Vegetable sculpture may amuse. It can evoke or express feelings. It can realize a concept, effect a therapy, astound, make money.

Within the food economy that it also extends, the new vegetable and fruit carving flourishes as part of an independent field of practice with its own evolving conventions and *raisons d'être*. A feature common to guides to vegetable sculpture is marketing. The manuals frequently include Web links, phone numbers, and paper order forms. The information encourages the apprentice to assume the role of consumer as well as adept. She can purchase specialized tools for carving, the services of a master carver, a culinary tour that includes carving lessons. Consumption, the participatory attitude, knowledge, and skill conjoin in the field of vegetable sculpture. The field features accomplished specialist carvers as producers, formal and informal systems of apprenticeship, competitions, commentators and judges, consumers to appreciate and also to pay for the creative work and technical expertise of specialists, and amateurs whose activities blur the line between producer and consumer.

Today's ephemeral vegetable sculpture moves outward from the dining room, and it assumes a permanent form through visual media. The new carving may function as a kind of über-garnish but to an entire table, buffet, or dining room, not merely to a plate. It is equally likely to be found in the context of a carving competition or at a venue decorated for an event whose primary focus may not be eating at all. Increasingly, the experience of vegetable carving – its visual consumption – takes place via a medium that is electronic. The specialized treatment of vegetable and fruit carving has largely taken its contours from published books. But the genre has additionally taken the usual leap across media to television, video, DVD, and the Internet. A few DVDs and videos give master demonstrations of fruit and vegetable carving. The Internet is now the best bet for seeing carving and for helping others to see it. The Internet hosts an ever-changing stream of formal and informal digital footage on this topic. Go online and see amateur and professional vegetable carvers in action, hear them discourse on their art, admire finished cantaloupe lanterns and taro roses panned in the round. In May 2008, the *New York Times* registered the arrival of fruit and vegetable carving on the food scene in the article 'Knife Skills.' The online version of the article was, naturally, accompanied by a video: 'The Art of the Watermelon' shows the genesis of a bust etched out of a watermelon. We have seen that the new vegetable carving does not flatter the stomach. The new media contexts further remove the base material, the process of creation, and the end result alike from the sphere of the culinary and gastronomic.

Instead of emphasizing taste and smell, the new vegetable carving flatters the distal sense of sight. Digital media conveniently make knowledge about vegetable carving widely and more permanently available. These media further the art of vegetable sculpture, while completing its dematerialization.

******

The rise of vegetable carving as a commercial practice and an art form coincides, in America, with anxiety about food resources and the sustainability of the global food system. The coincidence is telling. While delighting the eye and the imagination, vegetable sculpture shows us blind spots in the meat economy. The most basic equations of the food system are simple. In the food chain, American humans prefer meat, despite its excessive ecological cost. But meat is yoked to vegetable, to which it owes its existence. Vegetables not used as food do not feed people. The refusal to use vegetables as food implies that they are expendable, of little or no worth as such. Vegetable sculpture parallels other now-common cases of the appropriation of edibles for human consumption but not as food, such as the conversion of corn into ethanol fuel. The new vegetable sculpture fetishizes vegetables according to the tastes of a meat economy that largely overlooks the costs of its own propagation.

******

We have seen that vegetable carving has a long, if thin, history traceable in specialized print sources to Renaissance Europe. The *hautes cuisines* that the French system produced have probably always reserved a corner for the carved vegetable and fruit garnish. Today's complex vegetable sculpture also recalls other European predecessor traditions for food sculpture in ice, spun sugar, and pastry. Remarkably, contemporary American and European works on vegetable and fruit carving do not hark back to these Continental traditions. Rather than reclaim its own local history, the new vegetable carving has created a separate myth of exoticism to explain its own provenance.

Works on vegetable carving written for readers in America and Europe over the last 30 years describe their art as *ancient* and *Asian*. The works do not offer a wealth of historical detail. Any information about provenance appears in short, general introductions. Collectively, the new works on vegetable carving adopt a contradictory stance. They insist on exotic origins in Thailand or China, while withholding detail about these origins. Nor do they inform us about how and why the foreign custom passed into American and European cultures.

Vague rhetoric that emphasizes old and Oriental origins of vegetable carving serves several purposes. Foreign provenance explains why an American or European may be unaware of the practice. A passing reference to a Thai tradition or Chinese custom gives just enough information to intrigue. Such a reference may strike a chord for the reader – the advanced gastro-culinarian and aspiring vegetable sculptor in Europe or America – who has eaten in a Thai restaurant or traveled to Thailand, but who does not possess an intimate knowledge of the culture or customs or language. Antiquity and exoticism further justify to the same reader why she should be interested in learning about vegetable carving. But in the absence of precise or extensive information, ornate vegetable carving has little specific significance attached to it. It emerges from no particular context within some foreign culture. It arrives in America conveniently unburdened of cultural baggage, of ideological weight. Unspecified origins leave the reader-practitioner free to appropriate, naturalize, and enjoy vegetable carving as she sees fit. No information, no background, no history – no strings attached.

In fact, China and Thailand have longstanding traditions of ornate vegetable and fruit carving. Writers from these countries have contributed extensively to the recent literature. Japanese works occupy a less prominent place in the library available to American and European readers. It is worth noting, with historians and anthropologists, that these cuisines have long been predominantly vegetarian, and that vegetarian traditions such as among Chinese Buddhists have prevailed despite the increasing availability of meat. Sources on Chinese food culture consistently note the antiquity and continuity of practices such as artful cutting as an integral part of cooking. Artful cutting is fundamental to cooking and to meeting expectations for food to be properly prepared and appealing. If artfully prepared vegetables and fruits aesthetically enhance meals, they are also used as religious offerings. Chinese authors on carving do incorporate some of this narrative in their works. In *Zhongguo shi diao* [*Food Carving and Presentation in China*], Renqing Zhang identifies a Chinese aesthetics of design for food based on artful cutting that is 2000 years old. Zhang cites evidence such as paintings on silk showing ritual food offerings for tombs dating to the Han Dynasty (206 BCE–220 CE). Other writers such as Xiang Wang connect ornate vegetable and fruit carvings to Tang Dynasty (618–907) imperial banquets but also to middle class aesthetics and to Buddhist offerings in homes and temples.

Contemporary American and European materials on vegetable carving and sculpture hardly engage Chinese-language sources, much less scholarship in any language on the cuisines and cultures of these nations. Curiously, the Rosen brothers' *Culinary Carving and Plate Decorating* appears ambivalent, if not satirical, about the extent of its debt to a Chinese work. The stylized Roman typeface on the opening page evokes

Vegetable Carving: For Your Eyes Only

Chinese characters, while the work announces, in very small print, that it is 'based on' a Chinese work named as *Vegetable Fruit Cutting Carving & Dish Decoration*, published by chef Chang Wen Liang. In the Rosen brothers' book, as in most works from the European and American authors, the vegetable carving described is a Western hybrid construction, a virtual product of the global imagination. The loose fiction of ancient, Eastern origins for contemporary vegetable carving tames and sterilizes Chinese history and Chinese contemporary culture. The new Western vegetable carving keeps both at a distance. It does not consider the role of vegetable carving in traditions of vegetarianism and in religious contexts. At the same time, the new Western vegetable carving plays to the longing for sophistication and the exotic.

It is noteworthy that the global food crisis and the blossoming of vegetable carving coincide with the latest wave of anxiety in America about the country's geo-political viability. Anxious, we imagine the Far East, especially China, as a source of trouble. The popular list of noisome elements includes ravenous consumption. But the art of vegetable carving as it is constructed in the West reveals the fallacy of displacing this anxiety entirely onto locales such as the Middle Kingdom.

******

This essay has sought to reveal the interrelated aesthetics, ethics, and cultural politics at work in vegetable carving. Vegetable carving has become fashionable, occupying a niche market in the American food economy. Contemporary vegetable sculpture, I have argued, has striking features that connect it to current debates about sustainability and globalization. Vegetable carving displaces the mouth in favor of the eye as the site of consumption for food. Native European roots exist for vegetable carving, separately from a more vital set of Asian practices in China, Japan, and Thailand. Yet, as we have seen, today's European and American literature of carving prefers exotic, Eastern origins, a myth rather than a history. The urge to tame nature through culture, but also to control culture, is ancient and not limited to food. Digital media accentuate the dematerialization of food that characterizes vegetable sculpture, even as the current moral-ecological imperative would have us all eating lower on the food chain, with a vegetable-based diet. As it is, we consume vegetable sculpture at will. Our taste for vegetable sculpture entails aesthetics and play but also economic and political consequences. In America, ornate vegetable carving is a prime product and symptom of life in the meat market and in the consumer economy.

Bibliography

Abramson, Julia. 'Deciphering *La Vraye mettode de bien trencher les viandes* (1926).' pp. 11–26 in *Authenticity: Proceedings of the Oxford Symposium on Food and Cookery*. Editor Richard Hosking. Totnes: Prospect Books, 2006.

——. 'Grimod's Debt to Mercier and the Emergence of Gastronomic Writing Reconsidered.' *EMF: Studies in Early Modern France* 7 (2001): 141–62.

——. *Food Culture in France*. Westport, Connecticut: Greenwood Press, 2007.

——. 'Legitimacy and Nationalism in the *Almanach des Gourmands* (1803–12).' *Journal for Early Modern Cultural Studies* 3.2 (Fall/Winter 2003): 101–35.

Adrià, Ferran with Juli Soler and Albert Adrià. *A Day at el Bulli*. New York: Phaidon Press, 2008.

'The Art of the Watermelon: Carving a Portrait of James Beard' [online video]. *The New York Times* (Wednesday 14 May 2008), Dining Section. http://video.on.nytimes.com/?fr_story=a967930bdb916b47861f1ce4df76a22bf320a1a1/.

Ball, Jay. *You Too Can Create Stunning Watermelon Carvings*. Logan, Utah: JBall Publishing, 2006.

Bradsher, Keith. 'High Rice Cost Creating Fears of Unrest in Asia.' *The New York Times* (29 March 2008), World Business.

Chang, K.C., editor. *Food in Chinese Culture: Anthropological and Historical Perspectives*. New Haven: Yale University Press, 1977.

Child, Julia. *The Way to Cook Meat* [videocassette]. New York: Knopf Video Books, 1985.

Costa, Narahenapitage Sumith Premala De. *Edible Art: Tricks & Tools for Master Centerpieces from Carved Vegetables*. (Translation of *Essbare Tischdekorationen zum Selbermachen*, Heel Verlag, translator Edward Force.) Atglen, Pennsylvania: Schiffer Publishing Ltd., 2006.

Dunlop, Fuchsia. *Land of Plenty: Authentic Sichuan Recipes Personally Gathered in the Chinese Province of Sichuan*. New York: Norton, 2001.

——. *Revolutionary Chinese Cookbook: Recipes from Hunan Province*. New York: Norton, 2006.

——. *Shark's Fin and Sichuan Pepper: A Sweet-Sour Memoir of Eating in China*. New York: Norton, 2008.

L'École parfaite des officiers de bouche, contenant Le vray Maître-d'Hôtel, Le Grand Ecuyer-Tranchant, Le Sommelier Royal, Le Confiturier Royal, Le Cuisinier Royal, Et le Patissier Royal. Paris: Jean Ribou, 1662.

Elffers, Joost. *Play with Your Food*. New York: Stewart, Tabori, & Chang, 1997.

Elias, Norbert. *The Civilizing Process*. Editors Eric Dunning, Johan Goudsblom, and Stephen Mennell. Oxford: Blackwell, 2000.

Freyman, Saxton and Joost Elffers. *How are You Peeling? Food with Moods*. New York: Arthur A. Levine, 1999.

Gargone, John. *Food Art: Garnishing Made Easy*. Atglen, Pennsylvania: Schiffer Publishing Ltd., 2004.

Gibbons, Gail. *The Pumpkin Book*. New York: Holiday House, 1999.

Giegher, Matthia. *Li Tre Trattati*. Padua: Paolo Frambotto, 1639.

Grimod de la Reynière, Alexandre Balthazar Laurent. *Manuel des Amphitryons*. Paris: Capelle & Renand, 1808.

Haydock, Yukiko and Bob Haydock. *Japanese Garnishes: The Ancient Art of Mukimono*. New York: Holt, Rinehart and Winston, 1980.

——. *More Japanese Garnishes*. New York: Holt, Rinehart and Winston, 1983.

Hertzmann, Peter. *Knife Skills: A User's Manual*. New York: Norton, 2007.

Hongwiwat, Nidda, editor. *Vegetable and Fruit Carving* (1999). 2nd edition. Bangkok, Thailand: Sangdad Publishing Co., Ltd., 2002.

Jaunault, Frédéric. *Bouquets de fruits et légumes sculptés*. Clamecy, France: Chiron, 2007.

Keller, Thomas with Susie Heller, Michael Ruhlman, and Deborah Jones. *The French Laundry Cookbook*. New York: Artisan, 1999.

Liu, Dilin. *Metaphor, Culture, and Worldview: The Case of American English and the Chinese Language*. Lanham, Maryland: University Press of America, 2002.

Loha-Unchit, Kasma. *It Rains Fishes: Legends, Traditions, and the Joys of Thai Cooking*. San Francisco: Pomegranate Artbooks, 1995.

Vegetable Carving: For Your Eyes Only

Lye, Colleen. *America's Asia: Racial Form and American Literature, 1893–1945*. Princeton: Princeton University Press, 2005.

Mitchell, W.J.T. *Iconology: Image, Text, Ideology*. Chicago: The University of Chicago Press, 1986.

Murphy, Kate. 'Knife Skills: Creating Feasts for the Eyes.' *The New York Times* (14 May 2008), Dining Section. http://www.nytimes.com/2008/05/14/dining/14carve.html

Narain, Sumitra. *Vegetable Carving: The Ancient Thai Art*. (Media Transasia Thailand Ltd. Bangkok.) 2nd ed. Karnataka, India: Akapolinea-Bangalore, 2003.

Polatsek, Tammy. *Aristocratic Fruits: The Art of Transforming Fruits into Art*. Brooklyn: Aristocratic Party Design, 2003.

Poon, Joseph. *Life is Short, Cooking is Fun*. Philadelphia: Joseph Poon, Inc., 2005.

Rosen, Harvey and Jonathan S. Rosen. *Culinary Carving and Plate Decorating*. Elberon, New Jersey: International Culinary Consultants, 1997.

Ruhlman, Michael. *The Soul of a Chef: The Journey toward Perfection*. New York: Viking, 2000.

Said, Edward W. *Orientalism* (1978). New York: Vintage, 1994.

Sittitrai, Penpan. *Sinlapa kankæsalak phak læ phonlamai* (*The Art of Fruit and Vegetable Carving*). Bilingual edition. English translation by Maneerat Sittitrai Amornkool. Krungthep, Thailand: Rongphim læ Thampok Charoenphon, 1980.

Sringam, Roongfa and Martin Amada. *How to Carve Fruits and Vegetables the Thai Way* [videocassette]. Samui Island, Thailand: Samui Institute of Thai Culinary Arts (SITCA), 2002.

Tan, Yuci. *The Art of Food Sculpture: Designs and Techniques*. Atglen, Pennsylvania: Schiffer Publishing Ltd., 2002.

da Vinci, Leonardo. *Codex Atlanticus. Il Codice Atlantico di Leonardo da Vinci nella Biblioteca Ambrosiana di Milano*. Editor A. Marinoni. 12 vols. Florence, Italy: 1973–75.

Visser, Margaret. *The Rituals of Dinner*. New York: Penguin, 1991.

Vontet, Jacques. *L'Art de trancher la viande & toutes sortes de fruits, à la mode italienne & nouvellement à la françoise*. [Lyon, *c.* 1650?] Médiathèque du Pontiffroy, Metz, ms. 1579.

Wang, Xiang (artist) and Louise & Othmar Fassbind. *Das grosse Lehrbuch der Gemüse- & Früchte-Schnitzerei. / The complete manual of vegetable & fruit carving. / Le grand manuel de la sculpture des légumes & des fruits. / Il grande manuale dell'intaglio di verdura & frutta.* (1997). English translation by Jenifer Horlent; French by Georges Billig; Italian by Sandra Zindel; text on hand surgery by Dr. Regula Umbricht. 2nd. printing. Udlingenswil-Luzern, Switzerland: Andy Mannhart AG, 2000.

'The World Food Crisis' (editorial). *The New York Times* (10 April 2008).

Wu, Jinrui and Zili Wu. *Ru chu shui guo da quan: Fen lei, xuan gou, ru zhuan, zhuang shi* (*The Complete Book of Fruit*). Xianggang, China: Wan li ji gou, Yin shi tian di chu ban she: Fa xing zhe Wan li ji gou ying ye bu, 1995.

Zhang, Renqing. *Zhongguo shi diao* (*Food Carving and Presentation in China*). Beijing, China: Zhongguo shang ye chu ban she, 1994.

The War of the Vegetables
The Rise and Fall of the English Allotment Movement

Lesley Acton

The seeds of discontent

The seemingly innocent and harmless occupation of growing vegetables for personal consumption was born from an almost revolutionary movement played out in 'parochial battlefields.'[1] The rural labourers and yeoman farmers had become disenfranchised from the land, and coupled with a lack of employment opportunities, they grew hungry and disaffected. Allotments were seen as a way of alleviating the distress of the rural poor.

The misfortunes of the rural labourer have been largely attributed to the process of enclosure, particularly from 1730–1845. During that time, many labourers not only lost their land but also their jobs. This loss, combined with the poor harvests of 1794–6 and 1799–1801, and the Napoleonic wars, contributed to a serious decline in the living standards of the English proletariat, especially those engaged in agricultural work.[2] By the end of the eighteenth century, there was a large underclass of hungry, downtrodden, and disaffected people that needed help. Reformers, such as William Cobbett,[3] recognized that the welfare system (the Poor Law[4] and Speenhamland[5] system), was not an effective answer, and accordingly, they lobbied for land grants for the labouring poor. Thus the allotment movement was born.

By 1793, there was an extensive movement to provide the labouring classes with a portion of land. However, by 1805, interest in land provision for labourers receded, due to the improved harvests and more stable food prices. Consequently, the movement failed. In 1819, the Select Vestry Act (Poor Law amendments) allowed parishes to let up to 20 acres of land for use as allotments. Despite this Act, by 1829, there were only a total of 54 allotment sites recorded.[6] In 1830 there was a labourers' uprising and a series of riots ensued. These became known as the Captain Swing riots.[7] A series of parliamentary statutes followed and, by 1873, the allotment movement was well established with a recorded 242,542 sites covering 58,966 acres.[8]

The modern allotment movement began with the 1907 Smallholdings and Allotments Act. This Act differentiated between allotments cultivated with a spade (a garden allotment), and an allotment cultivated with a plough (a farm allotment).[9] What is commonly referred to today as an allotment is, technically, an allotment garden, as defined by the Allotments Act 1922, a parcel of land not exceeding a quarter of an acre (10 to 40 rods, 27.5 x 9 metres), and 'wholly or mainly cultivated by the occupier for the production of vegetable and fruit crops for the consumption by himself or his family' (Allotments Act 1922). A farm allotment is any area of land not exceeding two acres (or five acres if provided by a local authority), attached to a cottage or not, and held by a 'tenant under a landlord and cultivated as a farm or a garden or partly as a garden and partly as a farm' (Allotments Act 1922).

The early success

The success of the nineteenth-century allotment scheme was due in no small part to the establishment of the high-profile, active and effective Labourer's Friend Society (LFS).[10] The society was established in 1815 and was a leader in acquiring land and raising awareness of the allotment movement. The LFS produced the *Labourer's Friend Magazine* (1834–1884), which gave useful practical information such as setting out allotments and tenancy agreements and rules.[11]

The first allotment sites were, according to Burchardt, those of Thomas Estcourt at Long Newton, Shipton Moyne (on the Gloucester/Wiltshire border) and date from *circa* 1795.[12] There is, he acknowledges, some debate as to the origins of the first site, which he believes is due to making an insufficient distinction between a true allotment site[13] and a potato ground.[14]

Potato grounds were quite different from allotments and only used for growing potatoes and not for other crops.[15] Farmers let potato grounds to employees at a market rent, or as part of a labour contract, in lieu of wages.[16] Tied labour ensured the farmer had an affordable and guaranteed labour pool at harvest time, which was particularly important during the Napoleonic wars because of a shortage of manpower. With the return of the men from the Napoleonic wars, there was a glut of labourers and farmers no longer needed tied workers. Eventually, potato grounds became more important for the income they generated than as a labour pool.

The benefits of allotments were thought to be myriad.[17] They gave the labourer an opportunity to rise above the humiliating reliance upon Poor Law charity, and with it a new-found sense of self-worth which served to make him more satisfied and compliant. Nicholls believed:

> the possession of a quarter of an acre of garden-ground may, and often will, make to the labourer and to his family the difference between want and sufficiency, between privation and comfort, between a contented mind and the cheerful fulfilment of the duties of his station, and a mind soured, hardened, and dissatisfied, prepared to yield to vicious promptings, and to rush recklessly into breaches of the law.[18]

Proponents of allotments justified their provision and existence in many different ways. They believed, for example, it improved morals and reduced crime. Around the time of the Swing riots, rural crime rose to a point where landowners were desperate to find a way to alleviate it.[19] Having an allotment, which was highly liable to be forfeited if the holder behaved badly, was generally considered an effective crime deterrent,[20] especially for crimes committed while under the influence of alcohol.[21]

Other advantages of allotments were cited as better food and improved health, which not only benefited the labourers but also their employers. Improving the health, wellbeing and fitness of the workforce allowed the landed classes to save money by reducing their poor law tax.

Importantly, allotments were not provided as a form of charity. To this end a fair rent was charged, generally the same as that charged to a farmer for the same land. In that way, the self-respect of the labourer was upheld or promoted, and the social stigma of accepting charity was avoided. Indeed, because it was so important that allotments should not be perceived as a charitable institution, it was commonly written into the by-laws of each society that should the holder apply for poor relief then he was to forfeit his plot. There were other restrictions for which failure to comply could also result in the forfeiture of the plot, such as failure to attend 'some Place of public Worship at least once every Sunday.'[22]

Set against a backdrop of unemployment, low wages, and the prospect of ending up in the workhouse, the allotment became a refuge from the bleak hopelessness that many rural labouring families must have felt. Barring drought or disease, as long as the labourers 'paid the rent, kept to the conditions of the tenancy and provided adequate labour, manure or other fertilizer or seed' they could, at least, keep body and soul together, and themselves out of the workhouse.[23]

Despite all these perceived benefits, there was still opposition to the provision of sites, particularly from farmers who felt the sites reduced or negated their opportunity to exploit and subjugate their workers through their ties to the potato grounds.[24] They also argued that labourers would be too busy digging their own plots at harvest time to seek employment in the fields. Furthermore, dissenters argued it would enable labourers to steal crops and pass them off as their own by claiming they had grown them on their allotments. Nevertheless, despite the opposition, the movement to provide land for labourers took off. Today, some two hundred years after the movement began, allotments are very much a part of our landscape and heritage.

At its inception, the allotment movement was seen as a way of alleviating rural poverty. Eventually, the demand for land for domestic agriculture moved into towns and cities. Despite ongoing opposition from large landowners, legislation was passed compelling local authorities to provide land for use as allotments.[25] By 1895, there were an estimated half a million plots. World War I 'gave a tremendous impetus to the extension of the allotments system,'[26] with numbers rising from 600,000 in 1913 to 1.5 million by 1918. Many wartime plots were on land requisitioned under the Defence of the Realm Act 1916. With the end of hostilities, the requisitioned land had to be given up. Despite demand, particularly from ex-service men, by 1929 the number of plots had fallen to less than one million.[27] The decline in numbers continued until it was again temporarily reversed by the next national emergency, World War II. By that time, the *rus in urbe* was well advanced, with three out of four allotments in urban areas.[28]

Dig For Victory

> It is becoming increasingly obvious that, amongst the plots calculated to bring Nazism to its knees, those measuring 300 square yards are destined to play an important part.[29]

By 1938, Britain imported 'two-thirds of her total food supplies.'[30] With the threat of an impending war, all available shipping space was needed for the war effort. Accordingly, Britain was facing the possibility of food shortages at best and starvation at worst. In August 1939, government minister Dorman-Smith announced, 'We are launching a nationwide campaign to obtain recruits to the ranks of the food producers.'[31] The objective of this Grow More Food (GMF) campaign was to provide half a million more allotments, mainly in urban areas, to enable the recruits to grow vegetables and raise rabbits and chickens for additional food. The Grow More Food campaign was subtitled 'Dig for Victory' – a slogan that caught the public imagination and stuck. At that time, there were about 740,000 allotment plots in England and Wales. By the end of 1942, there were 1,400,000 plots and an unknown number of home gardens and 'unofficial' plots under cultivation.[32] In March 1944, the government estimated that domestic agriculture accounted for '10 per cent (some 3 million tons) of all food produced in this country.'[33] The exact number of allotments was not known because authorities only furnished broad particulars from 1939–43 inclusive instead of making full returns.

Land matters

At the outbreak of hostilities, local authorities were empowered to requisition unoccupied land without permission. It was estimated that there were 600,000 plots on requisitioned land occupied with permission of the owner,[34] and common land with permission of the Minister.[35] Slightly later, councils were authorized under the Defence Regulation 62A to convert parks, playing fields, and 'any other land in their possession' for use as allotments.[36] This move was not universally popular, and the president of the Middlesex Cricket Club remarked that sport was necessary for the welfare of the nation. The Royal Parks of London did set aside some land for use as allotments, although not their sporting pitches. Many sporting clubs, however, did turn land over to agricultural use, because of falling membership numbers due to conscription.[37]

Councils were, and indeed still are, under a statutory duty to provide allotment sites. As the need for more housing and the accompanying infrastructure pushed up the value of land, the allotment holder who, with very few exceptions, was merely a squatter, was 'chased from pillar to post.'[38] There was also a belief that councils were 'anxious to dispose of [their] allotments as they [were] a financial liability.'[39] Viola Williams, who worked for the Women's Farm and Garden Association during the war, believed that as soon as the war came to an end, allotments became 'the prime targets for house building,' as indeed they are today.[40] Consequently, the need for security of tenure was, and is, a fundamental campaigning issue for allotment societies.

The National Allotment Journal (NAJ) was the published voice of the National Union of Allotment Holders, and it frequently expressed concern over the tenure of plots. In September 1942, the journal reported that at its annual conference there 'were no less than eighteen resolutions tabled on the important subject of tenure showing how strongly the movement feels that the Government ought to make a serious attempt to deal with this vital issue.' However, even during the height of the Dig for Victory (DFV) campaign, there were reports of allotment holders losing their land, 23,166 square feet in this instance, 'because the Plessy Company want to build a canteen.'[41] Concern over the post-war tenure of land was voiced as early as March 1942, when the *Ilford Recorder* reported the Essex Federation of Allotment Holders was concerned that sites would be 'snatched for jerry-building immediately after the war.'[42] However, on 9 August 1945, the same paper was reporting that 60–90 per cent of the area's allotments were uncultivated (although this situation was hotly denied by the allotment users).

Nevertheless, the initial response to the DFV campaign was described as excellent, with more 'allotments awaiting new plot-holders than of unsatisfied demand for the land.'[43] The Ministry of Agriculture distributed 191,000 DFV posters and 1,750,000 allotment application leaflets. There were 'masses and masses of leaflets…the radio was full of instruction about doing everything possible…we had leaflets on everything under the sun, that would help us to produce more food of various sorts.'[44] However, because there was a shortage of paper, pamphlets were 'rather on the small side.'[45]

The home front

The home front was very much regarded as part of the war arena. Propaganda was intended to make the kitchen and the allotment into a battlefield.[46] Its 'recruits' were urged to go and defend and strengthen the home front 'with the spade and the hoe as emblem and armament.'[47] Food, according to Lord Woolton, the Minister for Food, was a 'munition of war.'[48]

In August 1940, Churchill lent his support to the campaign in words borrowed from his iconic speech 'The Few':

> [N]ever in the field of human conflict was so much owed by so many to so few. Here is one way in which millions can show they appreciate that debt. Let them make a personal contribution to the Dig for Victory campaign. They will be helping to ensure that our people have that last week's supply of food that may well be one of the decisive factors in our victory.[49]

Viola Williams believed that when Churchill took office, things changed, and people stopped growing flowers and took notice of the DFV campaign.[50] Householders turned flowerbeds and lawns into vegetable plots. Vegetables were planted between flowers where 'the decorative foliage of beet, carrots and parsley [was] quite a pretty addition.'[51]

Local authorities set up demonstration allotments in parks and recreation grounds. Even cinemas distributed leaflets and had displays of tools and produce.[52] Competitions were held to encourage people to join the DFV campaign. Prizes were often in the form of War Savings Certificates.

Specialists from the Ministry of Agriculture gave countrywide lectures, as did gardeners from the Royal Horticultural Society. ICI employed a group of five lecturers, each one assigned to a specific geographic region. They gave talks to allotment societies, schools and colleges promoting plant-protection products.[53]

War Agricultural Committees were very much involved with the DFV campaign. One employee remembered how 'we were given a van, a big demonstration van…and we used to go round to the villages and park on the village green and open up, and do demonstration or advisory work, and we were known as the circus…It was absolutely tremendous the surge of growing vegetables and fruit and the collection of anything wild that could be collected that could be made into a preserve or into a sweet or something.'[54]

By September 1940, only 13 months after the campaign had been launched, many people were left with excess produce and no advice on what to do with it. In response, the Ministry of Food announced an official plan to help in rural areas. The NAS, along with local community councils, the Women's Institute (WI) and horticultural organizations, was asked to make estimates of the surplus fruit and vegetables. Whereupon, they were to be taken to collecting centres for grading and consigned for delivery to the greengrocery trade. Produce was disposed of locally wherever possible. For example, the WI sold it on stalls at local markets at wholesale price. One gardener arranged to distribute her large surplus to recently arrived troops in her neighbourhood.[55] Another, who had 'sacrificed a large part of his garden to vegetable production,' decided to give away the produce to the wives of servicemen, local schools and hospitals.[56]

One of the more unusual suggestions to deal with excess produce came from André Simon, who suggested wholesalers and retailers should steam or simmer leftover vegetables and press them into moulds. Then the housewife would have a choice between fresh vegetables or a slab of day-old cooked vegetables.[57]

Despite the enthusiasm, not everyone was in a position to join the campaign. It was suggested that those without access to land might take over derelict gardens, or try growing mustard and cress indoors, and radishes, lettuces and tomatoes in window boxes.[58] The aged or infirm were often reliant upon charitable friends and neighbours to supply them with produce.

It was not just the public who were urged to join the DFV. All military units were asked to plant their own vegetables and use them as much as possible. *The Times* reported that RAF stations were continuing to Dig for Victory and that a 'great number of the aerodromes will be self-supporting in vegetables.'[59] Vegetables, mainly potatoes and carrots, were grown along the runways. One participant described the Army programme on DFV as 'irksome.'[60] She believed that everybody had the same experience. Those who did the most digging were posted away just as the fruits of their labours came up. 'So a terrible malaise about this developed…it went right through the army and possibly the other services too.'[61]

This malaise and eventual abandonment by many participants of the DFV campaign was not really surprising. The many years of unending queues for food, shortages of everything, and hardships that were suffered during the war years took their toll. Many people had simply had enough and lost interest in the campaign. '[W]henever an appeal has been made to the general public to take up allotments for patriotic reasons many of the new tenants who have come forward have surrendered their plots as soon as the emergency is over.'[62] Within 15 years of the end of WWII, the recorded number of plots had dropped from a wartime peak of about 1.5 million to around 800,000.

Allotments today

Ever since the foundation of the allotment movement, the provision of land for allotments has been an area of contention and political debate. On the one hand, proponents of the system believed that allotments, while not 'a universal and infallible remedy for all the ills that effect labourers, [were] highly efficacious for their alleviation.'[63] On the other hand, many landowners and farmers were against the granting of allotments. Eventually, the early dissenters were superseded by civic authorities who, for the most part, decided allotment land could be put to better use, usually as housing estates. Since the 1940s, the number of allotment plots has steadily declined. The most recent figure (1996) cites about 297,000 plots.[64]

Despite some early difficulties, the Grow More Food campaign was extremely successful. The campaign played a large part in ensuring that the population of the UK remained well fed and healthy during the war years. Today, we are, once again, facing food supply problems (this time because of climate change, peak oil and rising food prices). The fight for land for use in domestic agriculture continues. The right for land to grow vegetables for personal consumption is no longer a parochial war; instead, it is becoming a global one.

Notes

1. Archer 1997, 11.
2. Hammond 1948[1911]; Gash 1935.
3. Cobbett 1823.
4. Mingay 1997; Webb 1910, 100.
5. Neuman 1969, 319; Gash op. cit.
6. Burchardt 2002, 36.
7. Hobsbawm 1969; Thompson 1988, 343.
8. Burchardt 2002, 225.
9. Thorpe 15.
10. Burchardt 2002, 91.
11. Ibid 94.
12. Ibid 241.
13. Burchardt 2000, 670.
14. Moselle 1995.
15. Burchardt 2000, 2002.
16. Cobbett 1823 §77.
17. Hickey 1855, 67.
18. Nicholls 1846, 22.
19. Burchardt 2002, 186.
20. Anonymous 1823, 9.
21. Burchardt 2002, 186–197.
22. Crouch 1994, 53.
23. Burchardt 2002, 166.
24. G 1844, 9.
25. Thorpe 13 §29.
26. Ibid 16 §39.
27. Ibid 19 §48.
28. Fay 1942, 18.
29. Cook 1941, frontispiece.
30. Clark 1939, 24.
31. Wilt 2001, 189.
32. Thorpe 19 §48.
33. Thorpe 41 §112.
34. Thorpe 48 §133.
35. Cultivation of Lands (Allotments) Order 1939.
36. Thorpe 19 §48.
37. *The Times* March 13 1940, 11.
38. Fay 1942, 2.
39. Thorpe 124 §300.
40. Imperial War Museum Sound Archive #20321.
41. *Ilford Recorder* August 28 1941, 5.
42. Ibid March 26 1942, 3.
43. *The Times* December 7, 1939, 6.
44. Imperial War Museum Sound Archive #20321.
45. Ibid.
46. Bentley 1998.
47. *NAJ* op. cit., September 1941, 1.
48. *The Times* September 26 1940, 2.
49. *National Allotments Journal (NAJ)* Autumn 1940, 3.

50. Imperial War Museum Sound Archive #20321 .
51. Cook 1941, 9.
52. *The Times* February 14 1940, 5.
53. Imperial War Museum Sound Archive #19530.
54. Viola Williams op. cit.
55. *The Times* September 26 1940, 2.
56. *The Times* August 10 1940, 2.
57. *The Times* September 16, 1940, 5.
58. Cook 1941, 9.
59. *The Times* April 11 1942, 2.
60. Imperial War Museum Sound Archive #18001.
61. Ibid.
62. Thorpe 226 §575.
63. Hickey 1855, 67.
64. Harrison 2004.

Bibliography.

Anonymous. *Plan for the relief of the agricultural poor, with a view to diminishing the poor rates and give permanent employment to the labourer*; in Hume Tract 38. Wycombe: M.C. Morris, 1823.

Archer, John E. 'The Nineteenth-Century Allotment: Half an Acre and a Row'. *The Economic History Review* 50 (1):21–36, 1997.

Bentley, Amy. *Eating for victory: food rationing and the politics of domesticity*. Urbana: University of Illinois Press, 1998.

Burchardt, Jeremy. *The allotment movement in England 1793–1873*. London: Royal Historical Society, 2002.

Carter, Frederick William Pearson. *The Penguin Book of Food Growing, Storing and Cooking* [Penguin Books. Penguin Special. no. 90.]: Harmondsworth, New York: Allen Lane, 1941.

Churchill, Winston. [cited 12 May 2008. Available from http://www.winstonchurchill.org/i4a/pages/index.cfm?pageid=420.

Clark, Frederick Le Gros M. A., and Richard Morris Titmuss. *Our Food Problem. A study of national security* [Penguin Books. Penguin Special. no. 32.]: Harmondsworth, 1939.

Cobbett, W. *Cottage Economy*. London: J.M. Cobbett, 1823.

Cook, Raymond Arthur. *Plots against Hitler. A book for the North-Country victory digger*. Gateshead: Northumberland Press, 1941.

Crouch, David, and Colin Ward. *The allotment: its landscape and culture*. Nottingham: Mushroom, 1994.

Dowling, John E., and George Wald. 1958. 'Vitamin A Deficiency and Night Blindness'. *Proceedings of the National Academy of Sciences of the United States of America* 44 (7):648–661.

Fay, C. R. *The Allotment Movement in England and Wales*. London, 1942.

G. *Garden allotments: Letter to James Wood, Esq., Secretary to Labourers' Friend Society, Exeter Hall*. In Hume Tracts A1/1–21. London: A. Varnham, 1844.

Gash, N. 1935. 'Rural Unemployment, 1815–34'. *The Economic History Review* 6 (1):90–93.

Goebel, G.V. *Greg Goebel / In The Public Domain*. 01 May 2008 [cited 6 May 2008]. Available from http://www.vectorsite.net/index.html.

Hammond, J.L., & B. Hammond. *The village labourer*. London: Longmans, 1948 [1911].

Harrison, John. 2008. *Allotment History – A Brief History of Allotments in the UK* 2004 [cited 1 Feburary 2008]. Available from http://www.allotment.org.uk/articles/Allotment-History.php .

Hickey, William. *The agricultural labourer: viewed in his moral, intellectual, and physical conditions*. London: Groombridge and Sons, 1855.

Hobsbawm, E.J., and George F. E. Rudé. *Captain Swing*. London: Lawrence & Wishart, 1969.

Longmate, Norman. *How we lived then: a history of everyday life during the Second World War.* London: Pimlico, 2002.
Mingay, Gordon Edmund. *Parliamentary enclosure in England: an introduction to its causes, incidence, and impact, 1750–1850.* London: Longman, 1997.
Minns, Raynes. *Bombers and mash: the domestic front, 1939–45:* London: Virago, 1999.
Moselle, Boaz. 'Allotments, Enclosure, and Proletarianization in Early Nineteenth-Century Southern England'. *The Economic History Review* 48 (3):482–500. 1995.
Munsell, Hazel E. 'Vitamins and Their Occurrence in Foods'. *The Milbank Memorial Fund Quarterly* 18 (4): 311–344. 1940.
Neuman, Mark D. 1969. 'A Suggestion regarding the Origins of the Speenhamland Plan'. *The English Historical Review* 84 (331):317–322.
Nicholls, George, Sir. *On the condition of the agricultural labourer : with suggestions for its improvement. Journal of the Royal Agricultural Society of England.* Vol. VII, part 1. London: William Clowes and Son, 1846.
Patten, Marguerite. *Victory Cookbook, Nostalgic Food and Facts from 1940–1954.* London: Chancellor Press, 2002.
Thompson, F.M.L. *The rise of respectable society : a social history of Victorian Britain 1830–1900.* London: Fontana, 1988.
Thorpe, H. *Departmental Committee of Inquiry into Allotments.* Cmnd. 4166. London: HMSO, 1969.
Webb, Sidney, and Beatrice Webb. *English Poor Law Policy.* London: Longmans & Co., 1910.
Wilt, Alan F. *Food for war: agriculture and rearmament in Britain before the Second World War.* New York: Oxford University Press, 2001.

The First Scientific Defense of a Vegetarian Diet

Ken Albala

Throughout history vegetarian diets, variously defined, have been adopted as a matter of economic necessity or as a form of abstinence, to strengthen the soul by denying the body's physical demands. In the Western tradition there have been many who avoided flesh out of ethical concern for the welfare of animals and this remains a strong impetus in contemporary culture. Yet today, we are fully aware scientifically that it is perfectly possible to enjoy health on a purely vegetarian diet. This knowledge stems largely from early research into the nature of proteins, and awareness, after Justus von Leibig's research in the nineteenth century, that plants contain muscle-building compounds, if not a complete package of amino acids. Nonetheless, we have no problem conceiving of a healthy vegetarian body. To ancient, medieval and early-modern scientists and the lay public who understood their theories, this idea would have been impossible, primarily because meat, it was believed, is the most nutritious substance, the only food that is wholly converted into human flesh.

The purpose of this paper is to chart the emergence of physiological theories that first made a vegetarian diet 'good to think' rather than an intentional and normally religiously motivated form of self-mortification. Contrary to expectations, the defense of the vegetable diet originates not with modern physiological science, but the mechanist theories of the late seventeenth century, and in particular the use made of them by a University of Paris physician, Philippe Hecquet, in his *Traité des dispenses du Carême* of 1709.

The setting that prompted the composition of this treatise is deeply concerned with religious practice nonetheless, and in particular the state of Lenten 'fasts,' which as they were defined by the Catholic Church at the time demanded abstention from meat and meat products during the forty day period preceding Easter. However, 'dispensations' for various ingredients were fairly easy to obtain, normally with the payment of a fee, according to which logic greater benefit would accrue to the church through the charitable uses of this money than if the faithful were held to a strict enforcement of Lenten restrictions. In particular, dispensations were purchased since the late fifteenth century in the pontificate of Alexander VI for butter and dairy products and one could also in practice be relieved from the obligation to abstain from eggs. By the eighteenth century such dispensations were granted *en masse* by a local priest in a yearly ceremony, but technically each parishioner was required to explain precisely why he or she felt it necessary to consume these products and sometimes even meat itself. The most common excuse, according to Hecquet, was medical. By reason of age, infirmity, or the physical demands of particular strenuous professions, people were allowed to dispense with Lenten prohibitions. The more solicitous clerics would even demand a doctor's note explaining the condition.

In the minds of most people, supported by their physicians, both at some level appealing to standard humoral physiology, a vegetable diet was palpably dangerous and left the body susceptible to numerous maladies. According to Galenic theory, although significantly altered since classical antiquity, most fruits and vegetables, as well as fish (also allowed during Lent) are cold and moist foods that generate phlegmatic humors in the body. Thus they surfeit the constitution with cold and moist elements that naturally exacerbate the tendency to catch catarrh, rheums, and various ailments that today we would categorize as colds, coughs, and flu. This was a very real fear, and to the extent that individuals recognize the motions of their own bodies in terms that are dictated by their culture and its reigning medical orthodoxy, not at all unfounded. That is, as much as the body is socially constructed through history, people did indeed get sick in the winter and they accepted the cause as ambient cold and damp as well as a diet of cold and moist foods. It was these very foods, moreover, that were the foundation of the Lenten diet, in late winter normally before warmer weather and 'hotter' spring comestibles became available. There is therefore an inherent contradiction between Christian foodways and popularly understood humoral physiology. One commands you to eat vegetables and fish during Lent, the other insists that these are the worst possible foods for maintaining health in cold seasons. The only possible solution for health-concerned individuals was to purchase a dispensation. By the eighteenth century, this appears to have become routine.

Hecquet published his treatise in defense of Lent and a vegetable diet partly motivated by his own austere lifestyle and Jansenism, but equally by his position as rector of the medical faculty at one of the leading universities in Europe. There had been informal books defending abstinence, most notably the celebrated confession of Luigi Cornaro, who after serious illness limited his diet largely to vegetables and lived to a ripe old age. His work was further popularized in the dietary treatise of the Belgian Leonard Lessius, a Jesuit, though the call to abstinence in both these works could not be consistently defended from a medical perspective, or at least neither could persuasively recommend a vegetable diet per se. It was not until the gradual abandonment of humoral physiology that such a position was even tenable.

In the latter half of the sixteenth century and into the seventeenth century a serious challenge to Galenic medical orthodoxy was presented by the so-called iatro-chemical school following the work of Paracelsus and van Helmont, which reconceived of digestion as a process of fermentation rather than 'coction' as it had been imagined, the latter a type of cooking of food in the stomach by heat. Despite the fact that the chemical theories were gradually combined with standard humoral physiology, there was still no logical way to defend a vegetable diet since 'watery' foods were still seen as diluting the stomach acid, thus hampering the process of digestion and leading ultimately to a whole new variety of ailments stemming from the accumulation of tartar, the result of improperly processed food. Not until the emergence of a new digestive theory, or arguably a revival of the very ancient theory of Erasistratus, was the entire question of vegetarianism reopened. According to this school, the iatro-mechanical physiologists largely inspired by the work of Boerhaave, digestion is a physical grinding and mashing of food or, as they called it, 'trituration.' This can only be accomplished properly when the body's solid and liquid elements are in proper proportion, giving tensile strength to the fibers of the stomach as it processes food into a smooth creamy paste or chyle, which is ultimately converted into what they believed was the nutritive substance of the body: lymph. The stomach is thus a muscle, much like the heart, grinding food in the rhythm of systole and diastole, and completing the process of breaking down aliments begun by the teeth.

Thus watery foods do not in any way hamper the digestion, but lubricate the process, and are a necessary component of the final 'milky' chyle, or broken-down food, that nourishes the body. To paraphrase Claude Lévi-Strauss's famous dictum that before being good to eat, food must first be good to think, then vegetables for the first time became good to think according to scientific authority. For the first time, a scientist asserted that one could live in health eating only vegetables. But it is also clear that Hecquet was trying to convince a populace both lay and professional of his idea, since most peoples' conception of what was going on in their bodies was still largely humoral, though peppered with some chemistry. The details of his discussion reveal a thoroughly well-thought-out program, the first in fact to defend vegetarianism with scientific backing.

Hecquet opens his treatise by questioning the disjunction between the popular expression of physical conditions and the advances that had been made in medical theory, which had largely abandoned humoral therapeutics. Arguably such terms as a 'cold stomach' and a 'hot liver' are still encountered today as people express what they believe to be happening in their own body, which only attests to the remarkable explanatory power of the humoral system. Hecquet was trying to convince the general public that their anxieties were unfounded, and based on an entirely erroneous conception of physiology:

> On a severement proscrit les noms de chaud, de froid, de pituiteux, de bilieux dans la cure des maladies: on ne croit plus ces termes de bel usage, & ils ne sont plus que méprisables restes d'une physique surannée. Cependant que des alimens passent pour pituiteux, pour froid, pour bilieux, qu'on s'accuse d'un estomac refroidy, d'un foye chaud, d'un temperament pituiteux, c'en sera assez pour solliciter une dispense, & peut-estre pour l'obtenir.[1]

Hecquet squarely places the blame for this misunderstanding on physicians, who should know better, and who should be careful to correct popular errors in physiology. Namely, foods neither have the power to chill or heat the body, nor are human constitutions inherently prone to humorally-derived maladies. The implication is that physicians would be cutting into their revenues if they refused to indulge their patients' symptoms, thereby accepting fees for prescribing dispensations. What patient cares to be told his stomach is not in fact cold, and that fruits and vegetables will do him no harm? Such a patient seeks his doctor's note elsewhere. Thus patients and physicians were complicit in perpetuating an outmoded conception of the body, and as Hecquet insists, his goal is to disabuse the public of their ideas, and in a word, get them to eat their vegetables.

Nor were patients entirely disingenuous in seeking these dispensations in the first place. It is easy to imagine that many people, merely from dislike of a fish and vegetable diet, and perhaps with less than devout piety, purchased dispensations merely so they could indulge in their favorite foods, and concocted elaborate medical excuses for their need to eat meat. But if we take Hecquet's own perception of the situation, people really did fear the Lenten diet. That is, they had so internalized a humoral conception of the body, being deeply ingrained in this culture, that they did indeed suffer psychosomatic symptoms after being relegated to vegetables. Hecquet points out:

> Il est tres-peu de personnes que l'approche du Carême n'allarme, tous craignant alors pour leur santé & pour leur vie, comme si le jeusne & l'abstinence devoient abbréger leurs jours, ou avancer leur mort.[2]

It is here that Hecquet launches his campaign to convince people that everything they know is wrong, and in fact this book might also be considered the first popular full-frontal assault of a dietary system that had reigned for nearly two millennia. First, there was the popular misconception that children should be chubby. Clearly this prejudice stemmed from a fear of losing children in a dangerous world where infant mortality rates were still alarmingly high. But the effect of overfeeding children, Hecquet insists, lasts through life and forms habits difficult to dislodge. Through adulthood those who can afford it continue to overconsume. Furthermore, *l'embonpoint* is not a necessary condition of sound health. A frugal regimen is far more conducive to smooth functioning of the digestive system; eating less ultimately leads to better nutrition because food is more thoroughly processed and absorbed more efficiently.

This idea stems not from an ascetic frame of mind, but from a new physiological model. If the stomach is overloaded it fails to perfectly grind the mass of food into a uniform paste. The stomach works much like a mill grinding flour – when too much is added at once, the grains come out uneven. Only with a small mass can this smooth consistency be attained and only then can the body absorb the nutrients contained in the chyle.[3] Thus, eating less is more nourishing. Equally significant, those foods that are easiest to break down are more nourishing, which immediately suggests that meat with its tough fibers is not the ideal aliment. To judge the best foods, Hecquet suggests seeking those that are most easily ground down by the teeth, because the action of the stomach is much the same.

The First Scientific Defense of a Vegetarian Diet

> De là sans doute on apperçoit déja quelle sorte d'alimens est préferable à l'homme: ce ne seront pas les chairs des animaux, mais d'autres matieres qui auront plus de disposition à estre broyées & pétries pour mieux passer dans cette liqueur laiteuse qui doit faire le sang.[4]

It is in fact in this so-called 'milky liquor' that Hecquet discovers the ideal foods. Those foods that are most easily mashed into a smooth creamy consistency furnish the best material in the lymph to replenish the blood. It is then the tiniest particles of nutritive matter, reduced through a million triturations, or grindings, that pass through the pores in transpiration, another key feature of the mechanist physiology. If the body is nothing more than a machine, the proper consistency of food in every stage of digestion is crucial to its efficient processing and absorption, and finally transpiration, the final stage of the animal-economy. This theoretical construct leads Hecquet to a complete and utter reversal of standard nutritive theory. It is not those foods which are most similar to our own substance, meat as an analogue to flesh and wine to blood, which nourish most, but foods that break down easily, and which have a certain inherent sweetness or even lack of flavor. Once again, it was these latter foods stigmatized by humoral medicine as watery and phlegmatic that were most feared in previous centuries – watery vegetables in particular. Hecquet insists that these are the most nutritious foods; fruits and vegetables, roots and farinaceous starches are most easily converted to this milky fluid in the body, and are most appropriate for maintaining health.[5]

A humoralist would never have doubted that bread should be the foundation of a healthy diet, though the idea of *similia similibus nutruntur* could never quite accommodate bread logically into its schema; only meat is manifestly similar to and therefore converted into flesh. For the mechanist, however, starches provide the perfect material to furnish smooth chyle and lymph, as well as bland vegetables and mild fish – precisely what was ordered for Lent. These foods also constitute the primary diet of robust peoples around the world, solid empirical evidence that one need not eat meat.[6]

In practical terms, the diet proposed by Hecquet is diametrically opposed to what his readers, educated elites, would have been most familiar with. It is, strangely enough, essentially a peasant diet, and in a culture that invested such strongly negative associations with common, simply prepared food, it is nothing short of revolutionary. It is also, in many ways, similar to the diet proposed by Jean-Jacques Rousseau later in the century in his *Emile*, a simple and 'natural' diet based on vegetables. It cannot be by accident that the first specific food Hecquet discusses was the most universally reviled of lowly peasant foods: the fava bean.

> Car pour commencer par les légumes les plus vulgaires, il n'en est point qui ne soient tout-à-la-fois tres sains & nourissans. Les Féves, par examples, décriées aujourd'huy jusqu'à n'estre plus le rebut du beau monde, & la pâture de misérables, furent autrefois en honneur.[7]

The First Scientific Defense of a Vegetarian Diet

Among the ancients, that is, beans were an esteemed food. It is only in modern times and precisely because of the connection with poverty that most people avoid them. If they cause gas and feel heavy on the stomach, it is only because they are eaten too highly seasoned, which leads to an accidental fermentation of the stomach contents and distressful flatulence. On their own, however, and when properly prepared, they are easily broken down and easily converted into nutritive matter. Moreover, counter to all previous medical advice, as well as culinary fashion, he advises that dried beans are much more healthy than fresh.[8] The fermentation to which Hecquet refers is not the chemical digestion as conceived by a rival school of physiologists, but precisely the source of problems one associates with stomach grumbling (*borborigmes*), belching and intestinal gas. With this in mind, dried beans cause the least disruption because there is little volatile matter in them, unlike spicy seasonings and wine which tend to inflame the insides. Beans are basically bland, and therefore apt for human nutrition.

Hecquet also recognizes a similar prejudice toward cabbage, and frankly acknowledges that the only reason people avoid cabbage is because the poor eat it. To be acceptable on finer tables, among foods that denote status and distinction, it must be disguised, which is itself the root of all digestive problems people might have with cabbage:

> ...les pauvres s'en servent, c'en est assez pour rendre méprisable au reste du monde, qui n'en fait cas qu'après que la perniceuse industry d'un cuisiner, dautant plus dangereux qu'il sera plus habile, en aura sû déguiser le goût, & en dérober mème jusqu'aux apparences.[9]

Among the other lowly plant-based foods Hecquet promotes are oats; the accusations against them are poorly founded, for whole nations subsist on them. Root vegetables are also elevated, in particular salsify, turnips, beets, and even the topinambur or jerusalem artichoke from North America. Rice too is eminently useful, and among those foods most easily converted into a smooth paste and a pure lymphatic fluid. We need not be reminded that most Asians live on rice. It is interesting that like vegetarian and health diets of later generations, Hecquet promotes ingredients from foreign cultures, not as a threat to native foodways or as an exotic diversion, but as a purer and more natural form of nourishment, less corrupted by excessive refinement. Again, the similarity to Rousseau and nineteenth-century health reformers is striking, as is his promotion of green salads, unencumbered by recherché dressings but rather simple as it comes 'from the hands of nature.'[11]

It is impossible to determine the precise effect Hecquet's theories had upon the public. His book was immediately attacked by his Paris colleague Nicholas Andry, one of the iatro-chemists. This controversy is beyond the subject of this article, but it is certain that no other physician before Hecquet had dared to propose a vegetarian diet using scientific arguments of the time. After him, it became possible to think of a healthy vegetarian body.

Notes

1. Philippe Hecquet, *Traité des dispenses du Carême*. (Paris: François Fournier, 1709), fol. aii. 'We have severely proscribed the words hot, cold, phlegmatic and bilious in the cure of sickness; these are no longer considered terms of proper usage, and they are only disdained remnants of an outdated physiology. Meanwhile foods are still described as phlegmatic, cold, bilious, so that one mentions a chilled stomach, a hot liver, a phlegmatic complexion – this will be enough to request a dispensation, and perhaps to obtain it.'
2. Ibid, p. 1. 'There are very few persons who approach Lent without alarm, everyone still fears for his health and for life, as if fasting and abstinence could cut short their days, or advance their death.'
3. Ibid, p. 10.
4. Ibid. p. 16. 'Without doubt we perceive already that some kinds of foods are preferable to humans: these are not the flesh of animals, but of other materials that will be more easily disposed to be crushed and ground so they can more easily be transformed into this milky liquor which must be made into blood.'
5. Ibid. p. 25.
6. Ibid. p. 28–32.
7. Ibid. pp. 55. 'To begin with the most vulgar legumes, it is not only that they are at all times very healthy and nourishing. Fava beans, for example, reviled today to the point of being considered rubbish among the elite, and the food for the poor, were once honored.'
8. Ibid. p. 57.
9. Ibid. p. 98. '…the poor serve it, which is enough to make it loathed among the rest of the world, who only use it after the pernicious industry of a chef, all the more dangerous the more skilled, in knowing how to disguise flavor, and to dissemble to mere appearances.'
10. Rousseau, *Emile*, Translated by Allan Bloom (New York: Basic Books, 1979), p. 87.

Bibliography

Andry, Nicolas de Bois-Regard. *Le Régime du caresme, considéré par rapport à la nature du corps*. Paris: Jean Baptiste Coignard, 1710.

Brockliss, L.W.B. 'The Medico-Religious Universe of an Early 18th Century Parisian Doctor: The Case of Philippe Hecquet', in Roger French and Andrew Wear, eds. *Medical Revolution of the Seventeenth Century*. Cambridge: Cambridge University Press, 1989.

Debus, Allen. 2001 'The Debate Over Digestion' in *Chemistry and Medical Debate: Van Helmont to Boerhaave* (Science History Publications, 2001) pp. 154–163.

Hecquet, Philippe. *Traité des dispenses du Carême*. Paris: François Fournier, 1709. Reedited in 1710 and 1730 in a version completed in two volumes under the title *Traité des dispenses du Caresme*. Cologne: Roderique, 1741.

Rousseau, Jean-Jacques. *Emile*. Translated by Allan Bloom. New York: Basic Books, 1979.

The Roman Vegetable Garden

Joan P. Alcock

The Romans believed that gardens were a haven of peace to enjoy and promote spiritual renewal. These included a kitchen garden devoted to vegetables and fruit, although there might have been a separate orchard. The most popular garden of the Roman republic was a vegetable garden; in imperial times the most important was a flower garden (Lawson, 97). Even in towns most houses grew some vegetables, because produce was always fresh and brought to the table without cooking; there would also be economy in the use of wood fuel. According to Pliny (*Natural History* 19.19.57) gardens were women's responsibility and if uncared for indicated a neglectful mistress who had to rely on the butcher or the market for victuals.

On large villas and farms bailiffs supervised staff working on different parts of the garden. The person tending vegetables was known as an *olitor* from *olera* (vegetables) (Farrar, 160). Large and small establishments would have some slave labour to do manual work, but even the poorest household tried to grow some vegetables to be self-sufficient, as indicated in the poem *Moretum*, when the lowly peasant Simulus, having prepared his morning meal, goes to till his vegetable garden to have some surplus to sell as a cash crop (Kenney, 1984).

Flowers were sometimes grown in kitchen gardens to add decorative value. Virgil (*Georgics* 4.25) mentioned that a Corycian settler planted a vegetable garden, which included lilies, vervain, and poppies, the latter being useful for seeds used to decorate bread (Bowe, 48).

Tools and practice

Many tools used by the Romans have modern equivalents, for once a tool is deemed to be functional it is rarely changed. The Roman spade, an exception, was made of wood but had no grip handle, which was invented in the medieval period. The Roman spade had an iron sheath or shoe to give it some protection. Otherwise the tools (*ferramenta*) mentioned by Roman writers are the same. There were iron forks, four- or six-pointed rakes, hoes, two-pronged hoes and mattocks, an adze, pruning-hooks and billhooks and shears. These tools had wooden handles. The iron has survived, the wood has not. A clever invention, the *specularium*, had panels of transparent stone creating a kind of wheeled greenhouse. Columella (*De Re Rusticae* 11.3.52–53) said that this was used to supply cucumbers to the Emperor Tiberius, who liked them so much that he wanted them every day. Manure was transported in baskets or on hurdles as seen on mosaics at Vienne and Saint-Romain-en-Gal (France). Baskets were also used as containers for crops. Fertilizers included ash, ground-up pottery, and kitchen waste, as well as manure.

Columella (*De Re Rusticae* 2.15.6) and Pliny (*Natural History* 18.38.135; 18.37.137) recommended pulverized lupins and vetch.

After being dug over, beds were often divided into plots (*areae*); Simulus provided one plot for each vegetable. Pliny suggested that plots were given raised borders to keep in the water; paths should surround plots so that they could be easily weeded. Columella (*De Re Rusticae* 2.10.26) and Pliny (*Natural History* 19.20.60) confirmed this, recommending that gardeners, rather than trample on plants, should weed round each half of a bed in turn. The best size was 3 metres (10 feet) wide and 15 metres (50 feet) long so that beds could be watered from a path. The Roman villa at Boscoreale near Pompeii showed evidence of such beds with water being supplied from a nearby well. The Fishbourne villa (West Sussex) kitchen garden was adjacent to water supplied by a wooden pipe (Cunliffe, 107).

Pests

Control of pests and disease was essential if the best crop was to be achieved. Control might be possible by sowing one crop against another. Pliny (*Natural History* 19.58.179; 12.3.8) recommended bitter vetch to protect turnips, chickpeas to deter caterpillars from eating cabbages, and heliotrope to deter ants. He also suggested compounds or liquids of olive lees and soot, and juices of wormwood and horehound. Seeds soaked in cooked horehound juice might deter rodents. Virgil mentioned soaking beans in nitre or *amurca* before sowing, the intention being both for protection and to produce beans which were bigger and more easily cooked (Jermyn, 59). *Amurca* was a produce of manufactured olive oil, probably one of the vegetable resins that was skimmed and boiled down to one-third of its original volume. Columella (*De Re Rustica* 1.6.12–14) said that beans treated in this way were less vulnerable to attack from weevils. When they and other produce were stored in buildings with walls covered with a slime of *amurca* mixed with chaff, they were less likely to be eaten by mice and weevils.

The best protection was eternal vigilance, and slaves or the lower members of a household checked for aphids and greenfly. The last resort was superstition and religious rites. Palladius (*Opus Agriculturae* 1.123.860) even mentioned sending a barefoot, menstruating girl to walk three times round plants. Sacrifices were made to Robigus (the God of Rust). At the festival of *Robigalia* on 25 April (Ovid, *Fasti* 4.906–42) the entrails of a rust- or red-coloured dog were sacrificed in the hope of averting blight or mildew. More practical measures were to use rotation of crops or even leaving ground fallow for a year, although this might be difficult for many households.

Sources

The Romans were not short of horticultural manuals to give them advice. Cato (234–149 BC), Columella (active *c.* AD 60–65), Varro (116–27 BC), and Palladius (active fourth century AD), gave instructions about growing vegetables. Of Pliny the Elder's (AD 23–79) 37 books on natural history, several deal with plants, especially fruits and vegetables, which are discussed for their culinary and medicinal properties. In addition, such medical exponents as Galen (AD 129–199) detailed their remedial benefits, and the cookery book accredited to Apicius provided recipes. Anthimus (active AD 474–534) wrote a letter, *De Observatione Ciborum,* on dietetics, half-cookery book, and half-medical text to Theuderic, King of the Franks, which provides information referring to the later Roman Empire. Details relating to vegetables are mentioned in many texts, especially the poems of Horace (65–8 BC), Martial (AD 40–103), and Ovid (43 BC–AD 14), and the letters of Cicero (106–43 BC). Pliny the Younger (*Letters* 2.17) in his description of his villa at Laurentum mentioned a well-stocked kitchen garden. He was so proud of this that visitors were routed through it to get from the atrium to the main garden.

Archaeological evidence has revealed details of the layout and treatment of gardens in Pompeii and Ostia, on small farms and larger villas and in the variety of vegetables grown. Beds discovered outside the Balkerne Gate in Colchester are said to bear a resemblance to asparagus beds. Excavations at forts have provided evidence that soldiers grew a large variety of their own vegetables, which they could eat. The evidence includes details of supplies on writing tablets such as those found at the fort of Vindolanda in Northumberland (Bowman).

Illustrations of vegetables are found on mosaics and frescoes. Floor mosaics, especially those in North Africa, often show the remains of food, including vegetables scattered after meals. Wall mosaics reveal details of techniques or implements. Those from Vienne were part of a calendar which showed scenes of sowing seeds, gathering crops, grafting trees, and the taking-out of manure on a hurdle.

Produce

Vegetables grew wild in Europe and the Near East from prehistoric times. Northern European tribes gathered them as a welcome addition to diet, but it was the Greeks and Romans who realized the value of a cultivated crop and appreciated their use. Ofellus, Horace's neighbour, knew the simplicity on working days of eating smoked ham shank and greens (*Satires* 2.2.116). Roman writers treated vegetable growing seriously, emphasizing low cost of production, fresh flavour, and the fact that such cultivation gave great pleasure.

Vegetables were grown to supply a household but, as Varro constantly recommended, it was profitable near a city to have a large garden to sell any surplus in the market-place. If the household decided to keep its produce, some products like mushrooms might be dried; others like cabbage stalks, fennel, parsnips, and asparagus could be pickled by adding salt and then brine or vinegar.

Obviously the range of vegetables did not include those introduced into Europe in the sixteenth century from the New World, but the expansion of the Roman Empire meant that vegetables were brought from conquered regions to enhance the Roman table. Conversely the Romans introduced many new vegetables to parts of their conquered empire or improved the quality of those already indigenous to an area.

These in particular included root vegetables, which when obtained from the wild were usually hard and woody. In Britain evidence has shown that the Romans expanded the Celtic diet with such vegetables as parsnips, turnips, radishes, cabbages, carrots, celery, fennel, and cucumbers, and they increased the growing of carrots, beans, and peas (Alcock, 63–66).

Herbs, though not a vegetable, were vital in the preparation of food and a large variety was cultivated in gardens. These included bay, basil, chervil, dill, lovage, marjoram, mint, mustard, oregano, parsley, rue, sage, savory, and thyme. Apicius' recipes indicate the use to which they could be put. Many were incorporated into medicines. Herbs grew wild in the Middle East but spread from there by trade or human contact to the Mediterranean regions and to northern Europe. Herbs travel well because they can be dried. They are easily grown, but some, such as parsley, may be more difficult to strike. The Greeks used them extensively because fewer spices were available to them. The Romans often preferred to use spices but continued to cultivate herbs in garden plots ready for immediate use. Apicius used fresh herbs liberally. The Romans also introduced herbs to other areas – dill, mustard, marjoram, sage, rue, and a cultivated parsley (Alcock, 63).

Legumes

Broad beans or fava beans were a major crop for all social classes in the classical world. They were cultivated from a wild variety found in the Near East that became prolific in southern Europe during the prehistoric era. Beans could be eaten boiled, roasted, and raw in the pod, which would provide fibre. The Roman poet Martial (*Epigrams* 13.7) commented that 'if pale beans bubble for you in the red earthenware pot, you may decline the dinner of rich hosts.' Beans could be ground and added to bread flour, but as their skins were tough, beans were often skinned and split ready for thick soup, although classical authors warned that eating it could lead to flatulence. Pliny, while remarking on the nutritional and medicinal properties of beans, said that eating beans 'clouds the vision' (*Natural History* 18.30.117–118; 22.69.140–141) and it is known that beans can produce favism, a haemolytic disorder common among Mediterranean people. Apicius (5.6.4) mentioned beans from Baiae, which indicates that they were prized.

In both Greece and Rome beans had a role in ritual as they were offered to the gods and used in ceremonies. At the agricultural festival of the *Florialia* (April 27) beans and lupins were scattered amongst the spectators to ensure fertility (Scullard 1981, 110–111; Persius *Satires* 5.177). On the other hand, on the *Lemuria*, the festival to appease the spirits of the dead, which was held on 9, 11, and 13 May, the head of the household, walking barefoot through the house at midnight, spat out nine black beans, then scattered black beans over his shoulder as a ransom for the living members of the household so that they would not be affected by the spirits of the dead (Toynbee, 64; Ovid *Fasti* 5.419–493).

Chickpeas are among the oldest and most-used pulses of the Near East, dating back at least to 8000 BC in Palestine. They were first gathered from the wild but quickly became cultivated. The Greeks served them as part of *tragemata* (the last course at a dinner) eaten fresh, roasted, or dried; they were also served at *symposia*. Seasoned with oil and salt, they were used as a vegetable. Aristophanes (*The Peace* 1136) mocked the name, giving it a double meaning as 'glans penis' and noted its tendency to cause flatulence. Chickpeas were cheap to buy and made a good soup that was sold by streetsellers, conferring on it an inferior status. Martial (*Epigrams* 1.103) said a helping cost one *as*. Anthimus (66) warned that, when eaten raw, chickpeas caused violent flatulence, bad indigestion and diarrhoea, but they were a diuretic and hence good for the kidneys. Green chickpeas might be used as a laxative.

Fenugreek was regarded as a both a fodder plant and human food. It was sometimes eaten at the beginning of a meal with *liquamen*, as it was believed that this would act as a laxative. Its juice was mixed with honey to sweeten the taste. The seeds with their distinctive curry smell were used to flavour sauces in Rome. Pliny (*Natural History* 13.2.13) said that an unguent of fenugreek, olive oil, honey, cat-thyme and marjoram was the most celebrated of its kind in the time of Menander during the second century BC.

Lentils were a major part of diet in the ancient world. In Greece they were a staple food; in Rome less so, as the Romans believed they could render men indolent and lazy. Even so their constant consumption by peasants indicates that they probably constituted the principal food consumed by the lower classes in the Mediterranean civilizations. The main use for lentils was as a base for soup or pottage, which could be flavoured with spices or herbs. Anthimus (67) suggested they would taste better favoured with vinegar or Syrian sumach. Meat or fish added variety to flavouring and texture. Apicius (4.4.2) gave a recipe for barley soup with lentils and vegetables:

> Soak chickpeas, lentils and peas. Boil crushed grain with these. Drain then add oil and the following chopped greens: leeks, beets, and cabbage. Add dill, mallow and coriander. Pound fennel seeds, oregano, asafoetida, lovage, and blend these with *liquamen*. Pour this mixture over the vegetables and grain, heat and stir. Put chopped cooked cabbage leaves on top.

The pea was first domesticated in the Near East around 8000 BC. From there it spread to Egypt, then throughout the Mediterranean world, and was soon cultivated in northern Europe. It is a highly sustaining protein food and, like the bean, could be dried and stored, but when boiled was the basis of a nourishing soup. Peas were regarded with approval as not causing flatulence. When fresh, peas provided a sugar substitute.

The Romans used vetch as a cheap fodder crop, but in the wild it would inevitably be found growing with grain crops and was therefore ground with the grain for flour. In Greece and Rome vetch was only eaten in times of scarcity.

Lupines were a food and a decorative garden plant. The flat, yellow seeds must be boiled, as they are poisonous when eaten raw. As lupines were eaten in large quantities by the lower classes in Greece, the plant became associated with poverty and was noted for a tendency to cause flatulence. Pliny (*Natural History* 18.36.133–136) remarked that it was shared by men and hoofed quadrupeds, implying in Rome the poorer classes ate it.

Vegetables

Although Pliny (*Natural History* 19.43.145–151) said asparagus grew wild for anyone to gather, he noted that, of all the cultivated vegetables, asparagus needs the most diligent attention. He mentioned that it was one of the most beneficial foods to the stomach; if cumin was added it would disperse flatulence of the stomach and colon. The Romans realized its commercial value and Cato (*De Agricultura* 161) recommended that it should be planted in reed beds and covered with straw throughout the winter to survive any frost.

One of the most prolific vegetables was cabbage. It grew wild but the Romans promoted its cultivation throughout the Empire, regarding cabbages as having healthy properties. There were many different varieties; some having sprouts akin to Brussels sprouts. Cato (*De Agricultura* 156–158) classified the different kinds and spelled out all the virtues necessary to health, stating that cabbages could be eaten raw, cooked, or pickled in vinegar. He recommended that cabbage should be eaten both before and after a feast to combat drunkenness.

Cabbage had value both for nutrition and as a medicine; eating it was essential to good health and digestion, especially as an excellent laxative. Pliny (*Natural History* 19.41.136–144; 20.33.78–96) described 87 varieties together with their medicinal properties and said that it would be a long task to make a list of all the praises of the cabbage. Theophrastus (*Historia Plantarum* 7.4.4; 7.6.1–2) recommended that cabbage should be boiled in *nitron* to improve both flavour and colour. Pliny and Martial (*Epigrams* 13.17) both suggested adding a pinch of soda to keep it green. Juvenal (*Satires* 1.134), on the other hand, scorned cabbages, saying they were the food of the poor. Oribasius (*Medical Compilations* 4.4.1), quoting Mnesitheus of Cyzicus, gave a recipe for what may be considered a kind of vinaigrette coleslaw:

> Cut up the cabbage with a very sharp knife. Wash the pieces and drain off the water. Add sufficient quantity of rue, and coriander; sprinkle with honeyed vinegar and grate on top a small quantity of asafoetida.

Celery grew wild and was a smaller and more bitter plant than its modern equivalent. It began to be cultivated by the Romans, thus becoming widespread. Anthimus (55) said that it could be added to the preparation of all foods. Seeds, leaves, and stems were used for flavouring. As celery was related to parsley, the leaves were wound into a victor's wreath. Its medical properties included an ability to retain urine, which seems somewhat confusing as celery juice was also regarded as being a diuretic. It could also be applied to wounds and stiff joints.

The Romans used the shoots of fennel as a vegetable and the seeds as flavouring to food; its roots were used in medicine. Cato (*De Agricultura* 117) used fennel seeds as a dressing for olives:

> Bruise olives and throw them into water. Change the water often and when they are well soaked, press out and throw into vinegar; add oil and half a pound of salt to a modius of olives. Make a dressing of fennel and mastic (or pine resin) steeped in vinegar in a separate vessel. If you wish to mix them together they must be served at once. Press them out into an earthenware vessel and take them out with dry hands when you wish to serve them.

An unusual vegetable was samphire, found on seaside rocks, which was collected in spring both in the Mediterranean and in northern Europe. It was eaten raw or boiled or, if brined, could be made into a pickle, which would keep for over a year.

Root vegetables

The modern beetroot is a sixteenth-century propagation. The plant in the classical world had a smaller bulb than the modern large root. It was cultivated mainly for its leaves, which were often used to wrap other foods, much as vine leaves are used today. The leaves were also popular in a salad with lentils and beans. The roots were not popular, as they were regarded as a laxative. Martial (*Epigrams* 13.13) condemned them as insipid and a common noon meal for artisans. Oribasius (*Medical Compilations* 4.1.23), quoting from Galen, said that beet and lentil stew is a good food if salt and *liquamen* are added. He praised this dish as a laxative.

Carrots were thin and pale in colour and more popular than parsnips (Andrews, 1949). The leaves and roots of both plants were used for medicinal purposes. Pliny and Columella both mention wild and cultivated parsnips; the unopened flowers of the plant were dried and used as herbs. Pliny noted that they were a bitter vegetable even if the woody middle was removed, but this could be tempered with honeyed wine. Apicius (3.21) had three recipes, which may apply either to parsnips or carrots. One, which requires the vegetable to be served raw with salt, pure oil, and vinegar, must refer to carrots.

Leeks were grown in Mesopotamia at the beginning of the second millennium BC, in the Near East, and Mediterranean lands. The Romans thought that the best ones were imported from Egypt, and the next best were grown in Ostia. They are a labour-intensive crop because they need transplanting, but their constant growth means that they can be available for most of the year. Their flavour is equivalent to that of onions; in fact Apicius preferred them, for they appear constantly in his recipes. They are a versatile plant: the green tops can be eaten raw in a salad, the white bulb can be cooked as a vegetable, and when chopped both parts can be used as a seasoning. Leeks grew wild in northern Europe but the Romans introduced the cultivated variety. Pliny (*Natural History* 19.33.108–109; 20.21.44–49) mentioned its medicinal qualities, especially as an antidote to a variety of illnesses. He noted that Nero ate them on certain days in the month to improve his singing voice.

Like leeks, onions added flavour to a dish, but were less popular in Greece and Rome. In the wild, they were known to the Celts, but the Romans introduced a domesticated version to northern Europe. They are native to the Near East where they were cultivated in the third millennium BC. Strabo (*Geography* 16.2.29) knew a variety called Ascalon that got its name from the excellent onion market in that Palestinian town. Onions store well, so they could be kept through most of the winter. They also had other uses. Columella (*De Re Rusticae* 12.10.1) gave a recipe for pickled onions:

> First dry the onion in the sun, then cool in the shade. Arrange it in a pot with thyme or marjoram strewn underneath, and, after pouring in a liquid consisting of three parts vinegar and one of brine, put a bunch of marjoram on the top, so that the onion may be pressed down. When it has absorbed the liquid, let the vessel be filled up with a similar liquid.

In both Greece and Rome onions seemed to have been regarded more as poor man's food. Apicius made great use of them, but Horace included them in his poor man's diet of onions, pulses and pancakes. Pliny (*Natural History* 19.32.101–107; 20.20.39–42) gave varieties of them, those with the strongest taste coming from Africa, followed by those from Gaul. Because mature onions are mostly composed of water, they were placed in the moist category of vegetables, giving them numerous medical properties.

Shallots known as 'pearls,' a Marsumian variety, were presumably as small as a cocktail onion and similar to the Egyptian variety. Garlic was important in dishes round the Mediterranean lands, though more popular with the Greeks than with the Romans. Apicius made little use of garlic. The cloves were smaller than those today, many having 45 cloves to each bulb, which may have made each clove milder in flavour. Garlic adds relish to dishes but anyone eating it has smelly breath and, if they have overindulged, has a smelly body. Roman comedy writers made fun of garlic eaters. Anthimus (61) emphasized that it is useful for people on a journey, but it should be avoided by anyone with faulty kidneys. Palladius (*Opus Agriculturae* 12. 6) said that smoked garlic is good because it will keep for a year. Pliny *(Natural History* 19.34.114*)* said that the Greek writer Menander recommended eating roasted beetroot with or after eating garlic to neutralize the smell. Galen suggested that garlic could be boiled to remove its bitterness, although this could make it less efficacious.

The Roman knew a kind of horseradish, although they had little use for it (Pliny *Natural History* 19.26.82). The radish, now mainly a salad vegetable, was used for both its roots and leaves, although eating leaves was more from necessity than from choice. There were two main varieties. One was akin to the bulbous variety known today; the other was a black, woody variety. They were eaten raw as an aid to digestion or could be boiled for a vegetable. Radishes were cheap to buy and had a medicinal use, being particularly useful for dispersing phlegm, although they could cause flatulence. Horace (*Satires* 2.8.8–9) noted that radishes whetted a jaded appetite. Apicius' recipe (3.14) would do the trick: 'Serve radishes with a pepper sauce made by pounding pepper with *liquamen*.'

Turnips (Andrews) were grown as a standby for winter food. Pliny (*Natural History* 18.34.126–132), even though he thought turnips were a poor man's food, admitted that their 'utility surpasses that of any other plant'; even its leaves could be eaten. Columella, in particular, thought turnips a filling food for peasants. Classical writers recommended that turnips should be boiled twice, with the first lot of water poured away. As a white vegetable it had decorative uses because Roman cooks could stain it to pass it off as something else. Pliny said that it could be stained in at least six colours, the most popular being purple. Apicius (3.13.1) had one recipe for boiled turnips sprinkled with oil and vinegar, but he used the vegetable as a base for flavours:

> Boil the turnips. Drain. Pound together plenty of cumin and somewhat less rue and asafoetida. Add honey, vinegar, *liquamen, defrutum,* and a little oil. Bring to the boil and serve with the turnips.

Salad vegetables

Salad vegetables included lettuce, rocket, and endive. Martial (*Epigrams* 13.14) complained that lettuce, 'which was once eaten at the end of our grandsire's dinner, now ushers in our banquets,' but then (*Epigrams* 11.52.5) says that lettuce is most useful for relaxing the bowels. The Romans seem to have had an ambiguous attitude towards them. They thought that rocket, eaten in large quantities, aroused to intercourse and Pliny (*Natural History* 19.49.126) gave the recipe: 'pound three leaves of rocket with honey water.' But they also thought rocket could be counterbalanced by lettuce, which cooled the blood. Horace (*Satires* 2.8.8) mentioned that the Romans used them both mainly as hors d'oeuvres. The Romans introduced them to northern Europe, where they were appreciated.

Cress was prized for medicinal and aphrodisiac purposes. Pliny (*Natural History* 20.50.130) liked the Babylonian cress, saying that it sharpened the senses and cleared the vision. Endive was served cooked because of its bitter taste when raw. It was used for both culinary and medicinal purposes. Seeds of cucumbers have been found in numerous excavations. The Romans ate cucumbers cooked or raw, peeled or with the skin left on. Pliny (*Natural History* 19.23.64) and Columella described how they could be cultivated under movable frames.

Weeds

Weeds are worth mentioning because when gathered from the wild they served as vegetables especially for poorer people, were used to eke out food supplies (Frayn) and also provided medicinal remedies. Nettles were boiled before eating to remove the sting, and the water provided an antiseptic. The thread in the plant was used to make netting and cloth. Mustard was used as a vegetable when gathered from the wild. Apicius (4.2.7) made a patina, which seems to be a purée of wild herbs, spices, *liquamen*, vinegar, oil, black bryony, mustard plant, and cabbage, which could be added to fish fillets or chicken. Chickweed, rich in copper, added variety to diet and its extracted juice smoothed sores and itches. Dandelion, though bitter when eaten on its own, became palatable when mixed with other leaves. Marsh thistle, burdock, shepherd's purse, clover, and groundsel could all be made appetizing. Yarrow was boiled to make a drink. It is now known by the name 'wound healer' as the liquid has antiseptic qualities. St John's wort had pain-reducing properties. Good King Henry's nutritious leaves were a source of vitamins and iron. Its leaves could be used as a poultice for sores. Orach, used as an herb, could be prepared in the same way as spinach and was noted for its medicinal properties. Silverweed, prized for its roots, could be boiled, roasted or ground for a thick pottage. Sorrel was both a vegetable and a medicinal plant as it was highly useful as a laxative. Horehound, used to make a drink, was useful to combat a sore throat.

Bibliography

Alcock, J. P. *Food in Roman Britain*. Stroud: Tempus Publications, 2001.
Andrews, A.C. 'The Carrot as Food in the Classical Period.' *Classical Philology*, 44 1949, 182–196.
——. 'The Turnip as Food in the Classical Era.' *Classical Philology*, 53 1958, 131–152.
Bowe, P. *Gardens of the Roman World*. Los Angeles: Getty Publications, 2004.
Bowman, A. K. *Life and Letters on the Roman Frontier. Vindolanda and its People*. London: British Museum Press, 1994.
Cunliffe, B. 'Roman Gardens in Britain: A Review of the Evidence.' In Dumbarton Oaks Colloquium of Landscape Architecture, Vol. VII. *Ancient Roman Gardens*. Washington: Harvard University, 1981.
Ferrar, L. *Ancient Roman Gardens*. Stroud: Sutton Publishing, 1998.
Frayn, J. M. 'Wild and Cultivated Plants.' *Journal of Roman Studies*, 65, 1975, 32– 39.
Jermyn, L. S. A. 'Virgil's Agricultural Lore.' *Greece and Rome*, 18, 1944, 44–69.
Lawson, J. 'The Roman Garden.' *Greece and Rome*, 19, 1950, 97–103.
Scullard, H. H. *Festivals and Ceremonies of the Roman Republic*. London: Thames and Hudson, 1981.
Toynbee, J. M. C. *Death and Burial in the Roman World*. London: Thames and Hudson, 1971.

The Bitter – and Flatulent – Aphrodisiac: Synchrony and Diachrony of the Culinary Use of *Muscari Comosum* in Greece and Italy

Anthony F. Buccini

Introduction: *L'amore dell'amaro*[1]

While surely all of the world's cuisines exploit each of the five basic tastes – sweetness, sourness, saltiness, bitterness, and umami – there are noteworthy differences of preference between individual cultures. As someone who grew up in the United States but in a family that adhered relatively strictly to Old World and specifically southern Italian foodways, I have long been struck by differences of taste preferences between the mainstream cuisine of the US and my own family's cookery. Of these differences, perhaps the most striking one is the prominence of bitter foods in the southern Italian kitchen against their relative marginality in the American kitchen. Without doubt, this contrast is related to the two cuisines' contrasting orientations: on the one hand, toward traditional small-scale agriculture and horticulture, artisanal production and foraging and on the other hand, toward large-scale agriculture and industrial food processing, with an eye toward ease of preparation and mass-marketing through appeal to the gustatory common denominator. Though by no means a purely American phenomenon, this mass-market approach to food production and consumption was especially well developed in the United States in the course of the twentieth century and resulted in the disappearance in the mainstream cuisine of a host of once commonplace things, from artisanal bread to organ meats to fresh vegetables. And to whatever degree there were bitter items present in the northern European cuisines out of which general American cuisine grew, they were, it seems, all lost.

In a sense then, with the greater prosperity, the greater availability of food in general, there came something of an impoverishment of the range of tastes and an increased focus on the flavours that appeal most to, dare I say it, children. Now, in the period extending over the past, say, twenty-five years, a certain part of the American public has embraced a more eclectic approach to food. And with that, some items that are characterized by their bitterness have suddenly become surprisingly popular and even trendy. I cannot help but marvel at how the German bitter digestive, *Jägermeister*, so much like the *Amaro Lucano* that I grew up with, has now become an essential part of the inebriation rituals of college students throughout the country. A wonderfully bitter vegetable that, when I was growing up, could be found only amongst the 'old school' Italian families, *broccoli di rape*, is now a commonplace on restaurant menus and in supermarket chains; much the same can be said for radicchio and chicory. Even the humble dandelion, which – to the horror of my non-Italian childhood friends – we plucked ourselves from the yard to make salads and cooked dishes, is now widely sold, often with multiple varieties on offer.

Italians, like members of other American ethnic minorities, are proud of their traditional cuisine and happy to share it with others, though these positive feelings are also mixed with a sense of dismay and even horror at the ways in which our dishes are sometimes reinterpreted, bastardized, and debased. There is also for some of us, I must confess, a certain melancholy that arises from the realization that foodways that were long peculiar to us and thus important elements of our ethnic identity become, in the course of culinary miscegenation, little more than fleeting status-markers for conspicuous consumers: the mass-marketing of such items cannot help but contribute to the dissolution, for better or worse, of that ethnic identity.

One peculiarly – though not exclusively – southern Italian bitter food item, that has so far not yet been fully 'discovered' and subjected to commercialization in the world gourmet marketplace is the bulb of the *Muscari comosum* or tassel hyacinth. There are broadly speaking two groups of people who know these bulbs as a food: first, those with their cultural roots in parts of Italy, Greece, and Turkey who regard them as a traditional delicacy, and second, those who, as culinary scholars or adventurers, have taken an interest in those regions' cuisines or, as classical scholars, have investigated the frequent references to the bulbs in ancient Greek and Roman literature. Indeed, outside of the areas where *Muscari* bulbs are traditionally consumed, most of the attention they have received has been in connection with their use in antiquity and their contemporary use has been largely neglected. In the following pages I build on what is known of them from classical times and consider the modern names of the bulbs as a means of shedding light on the post-classical history of this bitter bulb.

Muscari comosum: the plant and its many names

There are some forty species of plants in the *Muscari* genus of the family *Hyacinthaceae*, order *Asparagales*, and thus they are distant relatives of the plants in the family *Alliaceae*, to which belong onions, leeks, and garlic. The *Muscari*, which are native to Eurasia and thrive in the Mediterranean region, are characterized by their bunches of blue and purple flowers. Two species of particular relevance here are *Muscari racemosum*, with its grape-like clusters of flowers, and *Muscari comosum*, with an additional tuft of sterile tassel-like flowers standing up above the cluster of fertile flowers. The English names are 'grape hyacinth' and 'tassel hyacinth' respectively but one finds the name grape hyacinth commonly applied broadly to all members of the genus; an older name one encounters for *Muscari comosum* is 'purse-tassel' and a further scientific designation is *Leopoldia comosa*. The edible bulbs of these hyacinths are reddish, small – roughly nut-sized – and mucilaginous, with a more or less pronounced bitter flavour. Though grape and tassel hyacinths can be and are cultivated, as a food source the bulbs have traditionally been foraged from where they grow naturally in wild contexts and in marginal spaces around farm fields and roads.

In classical texts, the edible hyacinth bulbs are referred to in Greek as *bolboí* (sing. *bolbós*). This term could also be used to refer to the bulbs of several other plants, including onions and garlic, but in many literary and scientific texts it is used without qualification in clear reference specifically to the *Muscari comosum* for which it was without doubt the primary name (Csapo, 116). The name *bolboí* has been maintained as the basic designation in both the modern Greek standard and in many regional varieties as well, though in some areas other designations are used. For example, on Crete the usual term is *askordoulákoi* (sing. *askordoúlakas*), a secondary form that is derived morphologically from the word for garlic, *skordo*, through addition of a complex suffix, presumably with diminutive and affective semantic value. Another such derived word appears in the dialect of Megara, on the isthmus between Attica and the Peloponnese, namely, *brouboúlia*, a suffixed form of *broúbes* 'wild edible plants' (cf. standard Greek *brouba* 'herb') (Syrkou 2006, 163). Another term, *koutsomamádes*, appears to be in use in some parts of the Peloponnese and perhaps also on Euboia, though its distribution remains unclear. *Koutsomamádes* seems to be of very restricted use; searches in dialect dictionaries and on the internet have met with little success. Several Greek speakers I consulted have, moreover, never heard the word and it makes little sense to them, though it sounds as if it may be a compound of the colloquial negative prefix *koutso-* 'lame, stunted, bad, etc.' and the word 'mothers,' though how such a term would be semantically linked to *Muscari* is not readily apparent.

The Greek word *bolboí* was itself possibly – and its use in reference to hyacinth bulbs was almost certainly – borrowed by the Romans into Latin and in classical texts in that language the plant is referred to exclusively as *bulbi* (sing. *bulbus*). Confusion regarding the actual plant indicated with this name by some authors (Dalby 1996, 244) and the frequent qualification of the name *bulbus* with, most especially, a reference to its putative Megaran origins – *bulbi Megarici* (Dalby 2000, 145; 2003, 63) inclines one to think that for some sectors of Roman society it was an uncommon or exotic item. This conclusion is supported by the fact that in modern Latin, a.k.a. Italian, the reflex of *bulbus* – *bulbo* – is not used in specific reference to edible hyacinth bulbs, either in the standard or in the dialects. But this is not to say that the *Muscari comosum* is or has not been widely known and consumed in Italy.

The Italian names for the *Muscari comosum* are numerous and deserve to be categorized. First, we can distinguish between learnèd names derived from the scientific designations for these plants – such as *muscaro, muscarino* – and popular names. The popular names can in turn be divided into those that refer semantically to the flowering portion of the plant and those that refer specifically to the bulb. Among the first group are: *giacinto delle vigne* or *delle viti* 'vine hyacinth', *giacinto dal pennacchio* 'plumed hyacinth', and the particularly euphonious *zazzeruto* 'long-haired'.

Muscari Comosum in Greece and Italy

Of those referring to the bulb, there is a full range of secondary forms based on the names of the edible bulbs of the genus *Allium*, appearing either as derived forms with affective suffixes or else with qualifying adjectives. Interestingly, these names often tend to be fairly restricted in geographical use and within one and the same region a surprising array of names can be encountered, including various forms derived from one or more of the names of the main members of the *Allium* genus. For example, just in the province of Catania (eastern Sicily), one finds in different locales the following names: a) from 'onion' – *cipudazza, cipudruzza, cipudduzza sarvaggia, cipudduzzu*; b) from 'garlic' – *agghioru niuru*; c) from 'leek' – *purrazzu*; in addition, there occurs a further name with at least two variants – *rubittuni, trubittuni* – which perhaps refers to the reddish colour of the bulbs; cf. *rubino* 'ruby' and the Nuorese dialect (Sardinia) form *arrubina*.[2] In general, names derived from the words for 'garlic' and 'leek' are fewer and less widely in use than ones which refer to 'onion' though in some broad areas, for example, parts of Sardinia, 'garlic' forms appear to be dominant. Of the 'onion'-derivatives, the most common formation corresponds to the standard *cipollaccio*, which is 'onion' with the pejorative suffix *–accio* attached (cf. *porrettaccio* 'leek' + *-accio*). But other suffixes are also used, including the augmentative, thus *cipollone*, and the diminutive, *cipolline*, the form used in my family's Campanian dialect. In addition to the derived forms there are also those which combine the word for 'onion' with a qualifying adjective or phrase: *cipolla canina* 'canine', *selvatica* 'wild', *di serpe* 'serpentine' – and, of course, there are many variants exhibiting distinctive dialectal phonological developments of the names and suffixes.[3]

There remains one important family of names for *Muscari comosum* to consider here, namely, standard Italian *lampascione* (pl. *lampascioni*), the best-known such name outside of Italy. In the region of Puglia, dialectal variants of this word are the dominant and perhaps exclusive appellation for edible hyacinth bulbs, with variation involving not just slightly differing dialectal phonological developments but also non-*lautgesetzlich* deformation of the initial consonant and reformations of the suffix; among the forms cited by Rohlfs (1955/1959) for just the dialects of the Salentine peninsula (the 'heel of the boot') are: *lampascioni, lampasciuni, ampascioni, ampasciulu, pampascione, pampasciulu, vampascione*. Related forms are found to the north throughout Puglia and further variants have also been reported from Campania to the west, e.g. *lampagione, lampascione, vampasciuolo* (Hammer et al., 237), though derivatives of the word for 'onion' are widespread in much of Campania (*cipollina, cipollaccio*), as well as in Basilicata (*cevoddine, cipuddënë*) and in Calabria (*lampascione* but also *cipujuzzu, cipullazza*) (Rohlfs, 1977). Further afield, in central and southern Sardinia, alongside 'garlic'- and 'onion'-derivatives one also finds forms in the *lampascione*-family: *lampajoni, lampajone, lampaone* (Rubattu).

Finally, there are two alloglot speech-communities in southern Italy whose names for *Muscari* bulbs can be noted. The Albanian-speaking Arbëreshë of Basilicata have borrowed an 'onion'-derived form – *çëpuljin* – from the neighbouring Lucanian dialects

(Pieroni et al., 174). In Puglia, the Greek (*Griku*) dialects of the Salentine peninsula employ a form – *lampaúne* – belonging to the *lampascione*-family of names used in the surrounding Pugliese dialects (Rohlfs 1964, 289).

We will consider below the etymology of the *lampascione*-family but only after we examine further the properties and historical uses of *Muscari comosum*.

Qui Veneris ostium quaerunt... 'those who seek the harbour of Venus'[4]

From the extant discussions of *bolboí/bulbi* in classical Greek and Roman texts, there are a number of things that we can infer about the consumption of *Muscari comosum* bulbs in antiquity.[5]

First, it seems clear that among the Greeks *bolboí* were a well-known and reasonably common food, though it is difficult to say how widely – both geographically and socially – they were consumed. That there was a range of usual preparations for them among the Greeks is indicated by Galen and the evidence he offers is corroborated in other Greek and Roman texts: 'There are many different recipes for them: they can be boiled in water, as I have said, elaborately seasoned dishes can be made with them, they can be served fried, and they are popularly baked in the ashes' (Grant, 150). Implied here is that in opposition to the elaborately seasoned dishes with boiled *bolboí*, there were also simply seasoned dishes of boiled (and possibly also raw) *bolboí*, as likely reflected in a recipe that appears in the Roman *Apicius*: 'Serve bulbs in oil, *liquamen,* vinegar, sprinkle with a little cumin' (Grocock & Grainger, 253). More elaborate seasonings for *bolboí* are mentioned by Philemon, which are in his opinion required to make palatable the otherwise 'poor and bitter' vegetable: 'Look, if you please, at the bulb, and see what lavish expense it requires to have its reputation – cheese, honey, sesame-seed, oil, onion, vinegar, silphium' (cited in Athenaeus vol. I, 281). Actual recipes of a more elaborate nature are offered in *Apicius*: one involves cooking the bulbs, then frying them in oil and dressing them in a sauce of thyme, pennyroyal, pepper, oregano, honey, a little vinegar and, optionally, *liquamen*; a second recipe calls for the bulbs being boiled and pressed into a pan and served with a sauce of thyme, oregano, honey, vinegar, *defrutum*, date, *liquamen*, and oil (Grocock & Grainger, 253–55). Fried bulbs could also receive the simple treatment of being dressed just with *oenogarum* (ibid).

A further, seemingly very humble dish that was apparently well known in classical Greece deserves special mention here, namely, *bolbophakê*, a soup made of tassel hyacinth bulbs and lentils. This soup was thought of as hearty, stick-to-the-bones fare, judging from a passage by Chrysippus, cited in Athenaeus (vol. II, 221): 'In the winter season, a bulb-and-lentil soup, oh me, oh my! For bulb-and-lentil soup is like ambrosia in the chilly cold.' *Bolbophakê* is mentioned in at least one other place, again in a citation in Athenaeus (vol. VI, 151), this time in an anecdote from Lynceus, where the soup is served, as Dalby (2003, 64) puts it, 'at a courtesan's establishment,' namely that of the particularly well-regarded and witty Gnathaena, and is accidentally spilled by the men who were drinking in her house.

What this anecdote calls to mind indirectly is the association in classical times of *bolboí* and sex, for the Greeks and the Romans both believed that tassel hyacinth bulbs had noteworthy aphrodisiac power. Indeed, most classical literary references to the bulbs are at least partly concerned with that aspect of them. Regarding *bolboí*, Galen says 'some men who fill up on their food feel quite clearly that they hold their semen and are keener for sex' (Grant, 150). Further citations on this topic appear densely in a passage on *bolboí* in the second book of Athenaeus' *Deipnosophists* (vol. I, 277–9):

a) Alexis, dwelling on the aphrodisiac properties of bulbs, says: 'Pinnas, crayfish, bulbs, snails, buccina, eggs, extremities, and all that. If anyone in love with a girl shall find any drugs more useful than these...'

b) from Xenarchus: 'That house perisheth whose master's fate it is to lose his virile powers... Impotent is that house, and even the bung-necked comrade of the goddess Deo, the earth-born bulb, so helpful to its friends when boiled, has no power to save it now...'

c) from Heracleides of Tarentum: 'Bulbs, snails, eggs, and the like are supposed to produce semen, not because they are filling, but because their very nature in the first instance has powers related in kind to semen.'

d) from Dilphilus: 'Although bulbs are not easy to digest, yet they are nourishing and wholesome; further, they are purgative, they dull the eyesight, and they rouse sexual desire.'

e) citing a proverb: 'A bulb will do you no good unless you have the qualities of a man.'

Several Roman writers also mention the aphrodisiac power of tassel hyacinth bulbs and there are two such references in verse, one from Ovid (*Art of Love* book II, lines 421–4) and the other from Columella, which refer specifically to the variety from Megara: 'Let hyacinths' fruitful seed from Megara come, which sharpen men's desires and fit them for the girls...' (vol. III, 15). Note too Pliny's comment in his passage on various kinds of bulbs: 'venerem maxime Megarici stimulant' (*Pliny* vol. VI, 63). In Petronius' *Satyricon* (p. 341), bulbs are invoked along with snails as treatments for impotence and Varro, cited in the section devoted to tassel hyacinth recipes in *Apicius*, indicates that they may have been routinely featured at wedding banquets: 'If I have said anything about bulbs, those who seek the harbour of Venus should have them cooked in water, and they can be served at dinner when a marriage takes place, but they can also be served with pine nuts or with the juice of rocket and pepper' (Grocock & Grainger, 255).

While the Roman comments on *bulbi* most often refer to their alleged aphrodisiac quality and, moreover, often refer to them in that context as 'Megaran,' it seems reasonable to suspect that they were to some degree at least regarded as an exotic and luxury item. Taking the Greek comments as a whole, however, there does not seem to be any corresponding special attitude toward the bulbs among the Hellenes. The aforementioned discussion of *bolboí* by Galen seems to imply that they were a common food and one can draw a similar inference from two additional comments on them cited in Athenaeus: a) Diocles of Carystus is quoted as saying 'Wild vegetables fit to boil are the beet, mallow, sorrel, nettle, orach, [*bolboí*], truffles and mushrooms' (vol. I, 267);[6] b) Plato, in portraying his new citizens at dinner in the second book of the *Republic* is quoted writing '... they will have a relish also, such as salt, of course, and olives, and cheese; and they will cook bulbs [*bolboùs*] and green vegetables, the sort of which they make boiled dishes in the country' (vol. II, 131). To the Greeks then, it seems that *bolboí* had more of a rustic association than an exotic one.

There is no question but that the wealthier and literate sectors of Roman society were in matters cultural strongly attracted to and influenced by the Greeks and, judging from the extant literature from antiquity, this was no less – and perhaps even more – the case regarding culinary interests and tastes than matters in other cultural spheres. This Hellenophilia they displayed, as Dalby says, by their 'employing Greek or eastern cooks, by importing and paying high prices for Greek and eastern delicacies, by assiduously transplanting Greek and eastern plant varieties, by adopting Greek names for foods and for finished dishes' (1996, 198). And besides the influence the Greeks exercised on the Romans in the culinary sphere, they also had a central influence regarding the related field of diet and medicine (cf. Dalby 2000, 122), and it is clear that tassel hyacinth bulbs were thought of as having other effects on health, positive and negative, beside that of aphrodisiac (Grant, 150).

With these matters in mind, it seems natural to surmise that the use of tassel hyacinth bulbs in Rome may have been just one of the innumerable instances of a high cultural import from Greece. But from this it would be wrong to conclude further that the bulbs themselves or their consumption or even their use as an aphrodisiac were genuinely or generally foreign to ancient Italy. On the contrary, one must first bear in mind that Greek influence in Italy was not merely a cultural phenomenon that belonged to the upper echelons of society. Rather, the Greek presence and influence was also significant for the broader population of Italy by means of the presence of Greek merchants, artisans, and slaves. Perhaps more important for the case at hand, one must also remember that in the centuries preceding the founding of Rome itself, the Greeks settled extensively along the coasts of southern Italy and Sicily as well. Insofar as those settlers were in the habit of consuming *bolboí* in Greece, they surely continued the practice upon arrival in the colonies of Magna Græcia, where the tassel hyacinth also naturally occurred in the wild.

Muscari Comosum in Greece and Italy

Viagra antiquitatis

Though tassel hyacinths are nowadays cultivated for their edible bulbs, they were and still are to a degree a wild food for which knowledgeable people forage, just as was the case in classical antiquity. Such wild foods have traditionally constituted an important nutritional supplement to the basic diet for rural populations and also a source of folk medicines, many of which have demonstrable pharmaceutical value. In the case of *Muscari comosum*, we note that the Roman physician Dioscorides (first century AD.) wrote that tassel hyacinth, boiled with barley meal and pig's fat, caused tumorous growths to suppurate and break up, which is remarkable in light of the fact that a tassel hyacinth extract is currently used in chemotherapy for cancer (Riddle, 100).

While *Muscari comosum* does have some real medicinal qualities, its most famous rôle as *viagra antiquitatis* seems – to this writer and life-long consumer of the bulbs – dubious. But there are good reasons why people believed the bulbs have aphrodisiac power. First, the bulbs do have a demonstrable diuretic effect, recognized and commented on by the ancients. Second, *Muscari* bulbs contain a mucilaginous liquid and its viscous quality was identified with that of semen, as can be seen from the quote above from Heraclides; other foods associated with viscous and gelatinous liquids – snails, eggs, cows' feet, etc. – were likewise thought to be sexually strengthening for men. Third, in the particular case of tassel hyacinth bulbs, there is a striking physical resemblance between them and surgically exposed testicles, given the basic shape of the bulb, their reddish colour and the relation of bulb to stem (resembling vaguely the vas deferens); add to this that the flower bulb is itself a generative organ and it seems clear that, in terms of the metonymy of magic (as opposed to scientific) thinking, the tassel hyacinth bulb is a vegetal analogue of the testicle and through consumption its force is transferred to the eater.

Lampascioni: torches, testicles and tomfools

Earlier we delayed offering an etymology for the Italian name for the tassel hyacinth, *lampascione*, though it is, in fact, remarkably straightforward at one level.

The form *lampascione* and all its dialectal variants can be traced back to a Late Latin *lampadio-lampadionis* (Rohlfs 1956, 285). From a phonological standpoint, the one point of interest is the development of the cluster [-dy-] which in the particular form *lampascione*, used now in standard Italian, shows a non-standard/non-Tuscan and specifically south-eastern development, proper to the dialects of Puglia and neighbouring parts of Basilicata and Campania. Clearly, standard Italian has borrowed a south-eastern form, which is not surprising, given the prominence of tassel hyacinths in Pugliese cuisine and the lack of any cognate of *lampascione* or any other unique word for the bulbs in a culinary context from Tuscany or elsewhere in central and northern Italy. Yet, note that there are dialectally *lautgesetzlich* forms attested for far-flung Sardinia, where [-dy-] develops not to [-š-] but to [-y-], e.g. *lampajoni* (*vide supra*). Whether these forms developed in Sardinia directly from *lampadio(-ne)* or were borrowed from the Neapolitan dialect area, they indicate the previous existence of a larger area in which the tassel hyacinth had a name of this family.

The earliest attestations of *lampadio* are in the Latin translations of the Greek medical texts of Oribasius. There were two separate translations, the first of which likely dates to the sixth century AD with the second having been made not much later; both were made at or for the Ostrogothic court in Ravenna (Mørland, 16). One form occurs in the translation of the Synopsis, namely, *lampadiones*, the 'proper' Latin form (Molinier, 11). In the translation of the Euporistes, however, we find a striking set of forms from the two translations: two forms are found in manuscripts of the older translation, the conservative *lampadiones* but also *lampajonis*, and from the corresponding passage in the younger translation we find *lampagionis* (p. 444); these forms clearly reflect stages in the dialectal developments of the [-dy-] cluster.

The etymology of Late Latin *lampadio* itself deserves consideration and again, the basic facts are straightforward: this form is a borrowing from the Greek *lampadion*, a diminutive of *lampàs* 'torch, oil-lamp, (metaphorically) the sun'. The semantic link of this word to the tassel hyacinth is obvious if one considers the appearance of the flower: the 'tassel' or shock of flowers that stand straight up from the top of the plant resemble in form the flames of a torch.[7] In other words, the name *lampadio* for tassel hyacinth surely was originally inspired by the flowering top of the plant, rather than the edible bulb.

Perhaps also relevant to the development of *lampadio* as name for the tassel hyacinth is the fact that in Latin it also served as a personal name; for example, *Lampadio* is the name of a slave in Plautus' comedy *Cistelleria* and it was also the name of one of the earliest Latin grammarians, C. Octavius Lampadio, commentator on Naevius, who himself wrote a play named for a character *Lampadio*. Could this moniker in later times have come to be thought of as a proverbial dunce's name? In any event, it is clear that at some point, the primary reference of *lampadio(-ne)* was switched from the flower to the bulb and so the word, at least in some dialects, was available for broader, figurative application, with a fortuitous coincidence of vegetal and genital references that renders it uniquely appropriate as a term of abuse meaning, roughly, 'imbecile'.[8]

Së magna nGrecia e purë mMagna Grecia

The development of the southern Italian name *lampadio/lampascione* etc. for the tassel hyacinth must be seen as a reflection of the popular use of the plant from classical times on. The original distribution seems to have corresponded to a high degree with the area in which the bulbs are still eaten today and that, in turn, corresponds to the broad area of southern Italy which in antiquity was comprised of the Greek colonies, mostly on the coast, and the neighbouring interior areas, where the local Italic peoples came under strong cultural influence from the Greeks early on. Though it would be unwise to conclude that consumption of the bulbs was strictly the result of Greek influence, the textual and linguistic evidence suggests strongly that the popularity of the tassel hyacinth in southern Italy reflects in good measure the deep cultural impact of Greece on the area.

Notes

1. Many thanks to Amy Dahlstrom, Giannis Bertakis, Catherine Chatzopoulos, Eleni Staraki, Aristidis Vouzakis.
2. For the names from Catania, see Lentini & Venza 2007 and the supplemental list of plant names: http://www.biomedcentral.com/content/supplementary/1746–4269–3–15–S2.pdf. For the Nuorese form, see Rubattu 2006 at: http://www.toninorubattu.it/ita/NU-A3(ar-az).htm; for a comprehensive list of Sardinian dialect forms, see the entry for 'cipollaccio': http://www.toninorubattu.it/ita/C2.htm.
3. For a useful list of Italian names sorted by region, see: http://www.dymi.gr/www.dymi.gr/eu_progr_grundtvig_recipes_italy_general_el.html. For other Italian names, with pictures and information about the plant, see: http://www.dipbot.unict.it/alimurgiche/scheda.aspx?i=24.
4. Varro apud Apicius, translation by Grocock & Grainger 2006, 255.
5. For a brief but excellent introduction to the topic with extensive references to the classical literature, see the entry for 'bulb' in Dalby 2003, 63–4.
6. Gulik translates *bolboí* incorrectly as 'iris-bulbs' here and in other passages.
7. http://www.losgazquez.com/blog/wp-content/uploads/2007/07/tassel-hyacinth.jpg.
8. *Lampascione* manifests the union of two themes of abuse: 1) *imbecille vegetale*: *citrullo* 'cucumber', *cucozza* 'zucchina', *testa di cavolo* 'cabbage-head', *testa di rapa* 'turnip-head'; 2) *imbecille genitale*: *coglione* 'testicle', *minchione* 'prick', *testa di cazzo* 'dick-head'. Armando Polito, on an Italian chat-site, makes this connection as well: http://forum.dialettando.com/forum_new/show_thread.lasso?mode=.

Bibliography

Loeb Classical Library texts: Cambridge, MA: Harvard University Press:
    Athenaeus: *The Deipnosophists* (7 volumes), Charles Gulick, trans., 1927–41.
    Columella: *On Agriculture* (3 volumes), E.S. Forster & Edward Heffner, eds. & trans. , 1968.
    Petronius: *Satyricon*, Michael Heseltine, trans., 1987.
    Pliny: *Natural History* (10 volumes), W.H.S. Jones, trans.
Csapo, Eric. 'Deep Ambivalence: Notes on a Greek Cockfight (Parts II–IV).' *Phoenix* 47 (1993): 115–24.
Dalby, Andrew. *Siren Feasts*. London: Routledge, 1996.
——. *Empire of Pleasures*. London: Routledge, 2000.
——. *Food in the Ancient World from A to Z*. London: Routledge, 2003.
Grant, Mark. *Galen on food and Diet*. London: Routledge, 2000.
Grocock, Christopher, & Sally Grainger. *Apicius: A Critical Edition*. Totnes: Prospect Books, 2006.
Hammer, Karl, Helmut Knüpffler & Pietro Perrino. 'Checklist of South Italian Cultivated Plants.' *Genetic Resources and Crop Evolution* 38 (1990): 191–310.
Lentini, Francesca, & Francesca Venza. 'Wild Food Plants of Popular Use in Sicily.' *Journal of Ethnobiology and Ethnomedicine* 3 (2007). On-line at: http://www.pubmedcentral.nih.gov/articlerender.fcgi?artid=1858679
Molinier, A., trans. *Œuvres d'Oribase*. Paris, 1876.
Mørland, Henning. *Orbasius latinus*. Oslo: A.W. Brøgger, 1940.
Pieroni, Andrea, et al. 'Ethnopharmacology of *liakra*.' *Journal of Ethnopharmacology* (2002): 165–85.
Riddle, John. 'Byzantine Commentaries on Dioscorides.' *Dumbarton Oaks Papers*, Vol. 38 (1984) (Symposium on Byzantine Medicine): 95–102
Rohlfs, Gerhard. *Vocabolario dei dialetti salentini (Terra d'Otranto)*. (Bayerische Akademie der Wissenschaften, Philosophische-historische Klasse, Abhandlungen – Neue Folge, Hefte 41/48/53). Munich, 1956/1959/1961.
——. *Lexicon Graecanicum Italiae Inferioris*. Tübingen: Max Niemeyer, 1966.
——. *Nuovo dizionario dialettale della Calabria*. Ravenna: Longo, 1977.
Rubattu, Antoninu. *Dizionario universale della lingua di Sardegna*. 2006. On-line at: http://www.toninorubattu.it/ita/DULS-SARDO-ITALIANO.htm & http://www.toninorubattu.it/ita/DULS-ITALIANO-SARDO.htm
Σύρκου, Αγγελική. *Το μεγαρικό γλωσσικό ιδίωμα. Λεξικογραφική μελέτη*. Athens: Nisos, 2006.

We Talked About the Aubergines: International Diplomacy and the Cretan Diet

Andrew Dalby

When the Greek prime minister, Eleftherios Venizelos, first visited London in 1912, his journey could be said to have quite different aims – and utterly different results – depending on one's point of view. If he came to agree on a peace treaty to end the First Balkan War, a treaty to be negotiated under British auspices, the whole thing was a failure. The Second Balkan War broke out two months after the pointless treaty was signed. If he came to meet some people who were going to be influential in Europe in the next few years, he did extremely well: he met Winston Churchill, sealed a friendship with his Romanian counterpart Tache Ionescu and an even more useful friendship with David Lloyd George. If he came to see how the Anglo-Greek community, wealthy and powerful as it was, might help in the future to promote Greece's interests, he did very well indeed. Pro-Greek propaganda flourished in Britain from 1912 onwards. And there was one other achievement of the journey that he didn't foresee during his five-day journey by steamer and train from Athens: it turned out that in London he would meet his future wife. They married nine years later. Long after his death, Helena Schilizzi – Helena Venizelos as she became – wrote a memoir in which she tells the sympathetic reader how little she knew about the world at that time – as a 37-year-old millionairess– and how uncomfortable she was in high society. In the newspapers, however, she had followed Venizelos's astonishing rise to power, she had read some of his speeches. She liked his politics; she was an admirer. So, in December 1912, at the dinner given by the Greek community in London, at the Criterion on Piccadilly Circus, when Helena Schilizzi and Eleftherios Venizelos first encountered one another across a cold buffet (I use the word in its catering sense), what did they talk about? 'We talked about the aubergines,' she reports; 'we talked about the *imambayıldı*.'

This paper begins and ends with vegetables, but (I must now admit) it will occasionally take a leap toward wilder foods – for example, the missionary diet to which Billy Hughes, prime minister of Australia, alluded at the Paris Peace Conference in 1919. Occasionally it will stray beyond all forms of food, to the thorny problem of the Queen of Romania's pink chemise. But we will return *à nos moutons*, to our vegetables.

It was suggested by work I have been doing on Venizelos as Greek delegate to the Paris Peace Conference in 1919, the conference that hammered out the Treaty of Versailles and several other treaties that ended the First World War and, some might say, helped to bring about other wars. This work will appear in print under the title *Eleftherios Venizelos: Greece*, published in 2009 by Haus Publishing in their series *Makers of the Modern World.* But this paper was really inspired by that little detail in Helena's memoir, 'We talked about the aubergines.' It reminded me, as Lord Byron and (more recently) Margaret Visser have reminded us, that 'happiness ... much depends on dinner.'

The Cretan diet

Venizelos, though prime minister of Greece, was not born in Greece. He was born in Crete, in a village close to Chania; by ancestry he may well have been a Greek, but by birth he was a Cretan from the period when Crete was part of the Ottoman Empire. Athens, to Cretans, was officially the capital of a foreign country, though many Cretans longed for this to change. Chania (then the capital of Crete) was where he had his early career as lawyer, journalist, and local politician.

As a Cretan, he had flourished, like it or not, on the diet that is now the envy of Europe. Let me briefly characterize it. It features local olive oil, of which large quantities are produced very close to Chania; local wine, which is good throughout Crete and very good from the Kissamos district not far west of Chania; fish and other seafood; not much meat (and what meat there is has very little fat); snails from time to time; no shortage of bread; local fruit. Venizelos was born in Mournies, one of the ten villages around Chania where oranges will grow. I am sure there are ten villages, because the man who sold me orange juice in Chania said so. Then he tried to list them for me, but the list always came out at nine or eleven, as often happens with traditional lists. Whatever the true number, oranges grow at Mournies.

Above all, the Cretan diet contains vegetables in quite large proportion and in unusual variety. Tomatoes, cucumbers, courgettes, and – perhaps more important – wild mountain greens. To give a few examples: wild chicories, such as the species called *stamnagáthi* and *radikostiváda*, *Cichorium spinosum*, spiny chicory, not quite as painful as it sounds; Myrsini Lambraki in *Ta khorta* gives a good recipe for spiny chicory with kid. Bitterer still, *stravóxylo*, *Scabiosa cretica*, Cretan scabious, one of the herbs with interesting antioxidant qualities. Wild dandelions such as *pikrafáka*, *Taraxacum gymnanthum*, good in fresh salads and known for at least two thousand years as a medicinal herb. Wild purslane, *glistrída* or *chirovótano*, *Portulaca oleracea*, with a pleasant crunchy texture, the best green vegetable source of omega-3 oils. Mallow, *molócha* or *abelócha*, *Malva* spp., a good digestive, excellent with chicken; rocket, *róka* or *arómatos*, *Eruca sativa*, said by ancient authors to be the antidote to lettuce; and *volví*, of course, *Muscari comosum* (L.) Miller, requiring long baking, best served in an oily sauce.

That was the diet that Venizelos gave up when he moved to Athens – because, even in 1910, one did not eat like that in Athens. That was the food from which he was separated by many hundreds of miles when he spent 1919 at the Peace Conference in Paris.

International Diplomacy and the Cretan Diet

Vegetables and others at Paris in 1919

Each of the national delegations in Paris in 1919 took over a hotel, or certain floors of a hotel. The Greeks had the Hôtel Mercedes, just behind the Arc de Triomphe. The British had the enormous Majestic Hôtel, just a couple of minutes' walk away, although Lloyd George and his foreign minister, Arthur Balfour, were comfortably tucked away in two flats in the rue Nitot, lent to Lloyd George by 'a lady.' The Americans had the spectacular Hôtel de Crillon, an eighteenth-century palace that closes off the northern side of the Place de la Concorde. Magnificently converted in 1911, it was Paris's newest luxury hotel, and it put Harold Nicolson, one of the younger British diplomats, in mind of a battleship. Breakfast at the Hôtel de Crillon (I now quote the report of Charles Seymour, one of the experts in the American delegation): 'Breakfast turned out to be about the best in our experience – bread very nearly white and with the finest crust, the most perfectly fried sole I ever tasted, plenty of delicious butter and all the sugar we wanted. The coffee has a good deal of chicory in it.'

One notices the emphasis sliding like a butter-knife from quality ('the most perfectly fried sole') to quantity ('all the sugar we wanted'), and one might put it down to the greed of the university men in the American delegation. It's more complicated. That is the way things were in 1919, especially in Europe. Not far from Paris there were a great many people with not enough to eat. Austria and Hungary were visited by a delegation led by the South African General Smuts. Having been invited to a luxurious lunch in Vienna they were shocked by the poverty and hunger they saw afterwards in the streets, and Smuts vowed there would be no more such entertainments. 'His eyes when angry are like steel rods. But it was a good luncheon all the same,' said Nicolson guiltily. In Budapest no one except the delegation, not even Bela Kun's communist government, had enough to eat. A fact-finding mission to Moscow had a similar experience: 'piles and piles of caviar ... caviar and black bread and tea ... but we never had a meal outside our own house, and the Russians were frequently at our table.'

There were shortages even in Paris. When James T. Shotwell – 'short, with square shoulders, a straight back, and a full, forceful moustache the size of half a cigar' – arrived in Paris as the official librarian to the American delegation, he went to have tea at the studio of the expatriate painter Frank Edwin Scott. The two of them had to eat johnny-cakes and syrup, because there was no bread in some parts of Paris. Johnny-cakes are well known to the *Oxford English Dictionary*, though they weren't previously known to me: 'A cake made of maize-meal, in the Southern [United] States toasted before a fire, elsewhere usually baked in a pan.' To these observers, unexpected vegetables are evidence of privation – the maize meal that takes the place of bread; the chicory that substitutes for coffee – and it's something they were really concerned about. And yet when young Charles Seymour escaped to Brussels for a few days' holiday, in summer 1919, his attention seemed to focus once more on his breakfast. 'For breakfast Monday morning we had strawberries with an enormous bowl of thick whipped cream, an enormous omelet, crescent rolls, and all the butter and sugar we wanted for five francs.'

International Diplomacy and the Cretan Diet

Food shortages in the world beyond the conference were noticed again when on 27 February T. E. Lawrence and Gertrude Bell lunched with President Wilson's interpreter, Stephen Bonsal. In the dining room at the Crillon the vegetables were tender and the salad lightly dressed, but 'it was not a gay affair,' Bonsal reports. 'Lawrence, like all paladins, is high strung and has his moments of deep discouragement, and this was one of them ... "As for myself," he said, "I would like to retire to a little cottage, say in Somerset, and write a book about the rise and fall of the Abbasid Caliphate. It would abound in topical references and I would probably starve to death while doing it, just as so many other more deserving men are starving today."'

For those at the Peace Conference food never ran short, though the food at committee meetings was stereotyped. Charles Seymour in one of his letters home reported on 'tea at 4.30 p.m. at the Quai d'Orsay, at which you met some unexpected and eminent people.' 'To the adjoining banqueting hall we adjourn for a few minutes,' Harold Nicolson confirms, 'for tea, brioches and macaroons. The tea-urn gutters in the draught.' After the meetings came the great innovation of Paris in 1919, the cocktail parties 'from which' (as Lord Vansittart remembered, many years later) 'women barely tore themselves away in time to undress for dinner.' A conversation at one such party is reported by Frances Stevenson, Lloyd George's secretary and mistress. Marie, Queen of Romania, was telling Mr. Balfour about her latest shopping spree in Paris and her new pink chemise when she remembered that she was about to meet President Wilson for the first time. '"What shall I talk to him about, the League of Nations or my pink chemise?" "Begin with the League of Nations, and finish up with the pink chemise," Balfour advised her. "If you were talking to Mr Lloyd George you could begin with the pink chemise."'

Stephen Bonsal makes it clear that tea at the Quai d'Orsay and dinner at the Crillon were not his personal preference. On one occasion he went out to what he only identifies as 'a tourist hotel' for lunch with Essad Pasha, representative of the Muslim Albanians. 'It was a strange repast ... Behind his chair and also behind mine stood heavily armed servingmen, their blue tunics bulging out with pistols and their gorgeous belts bristling with yataghans ... Now and again, vexed by a tough morsel on his plate, Essad would drop his ineffective fork and, picking up the hunk with his fingers, would throw it disdainfully over his shoulder. The servingman never failed to catch it in his mouth and seemed to enjoy his share of the feast ... I was out of practice, so when the repast was over I simply slipped a twenty-franc note to the guardian behind my chair and he seemed to be perfectly satisfied. Evidently Paris had corrupted him as it has so many others.'

It is under the same heading of not-quite-conference food that I mention the discussion between Woodrow Wilson and Billy Hughes, the Australian prime minister, about Australia's intention to annex North East New Guinea. Wilson was a little shocked by the argument, which seemed to be all about what Australia wanted: self-determination was scarcely mentioned. Surely, Wilson shouted (it was necessary to shout because Hughes was stone deaf), surely Hughes didn't mean Australia to have New Guinea if the whole world was against it? Hughes listened, his hand cupped to his ear, and shouted back 'Yes, that's about it!' Wilson went as far as to concede that there could be a League of Nations mandate if the New Guinea natives voted for it. Hughes was amused by this idea. 'Do you know, Mr President, that these natives eat one another?' At that point Lloyd George, no doubt bursting with suppressed laughter – he loved Hughes's negotiating style – decided to move the discussion along by reminding president Wilson that under Australian tutelage New Guinea would lie open to Christianity: 'And you would allow the natives to have access to the missionaries, Mr Hughes?' 'Indeed I would, sir, for there are many days when these poor devils don't get half enough missionaries to eat.'

There was plenty of social life in Paris in 1919: weekends at Fontainebleau, picnic lunches at Versailles. During the week, Charles Seymour admitted, 'almost every day we get off for a walk, when it is not too hot, and stop for an ice or a drink at a café or terrasse. We have been having dinner parties up in our room for various people, and enjoying the long sunset from our balcony.' From his balcony at the Hôtel de Crillon Seymour would be looking west along the Champs-Elysées towards the butter-coloured sunset, and one can imagine him hoping the Peace Conference would go on for ever.

What Venizelos really wanted

What of Venizelos? Was he also hoping the Peace Conference would go on for ever? I can begin to answer this by quoting Harold Nicolson, a Venizelos supporter. Twice Nicolson took him to a party at the dandy Boni de Castellane's pied-à-terre, just round the corner from the Quai d'Orsay, 'where bishops and statesmen, generals and philosophers, assembled in a sixteenth-century atmosphere to listen to seventeenth-century music struggling against twentieth-century conversation.' On the second such occasion, when Nicolson and Venizelos had had enough of conversation and Cognac, they shared a cab back to their hotels. He was 'looking ill and tired,' Nicolson wrote. There were months of Paris still ahead of them: frustrating, probably inconclusive, and if a conclusion were ever reached it would surely be the wrong one.

This international lifestyle was not what Venizelos was used to. He was at ease with the travelling: people who live on Greek islands are used to travelling. He was accustomed to living away from home: as a Greek politician, which he rather suddenly became in 1910 at the age of 46, he was already an exile with very little money, at first living out of a suitcase in half a room in an Athens hotel.

The question that interests me is this. Assuming that he had money or that someone else was paying for his meals, did he look forward to the luxury of the diplomatic circuit? Did he long, like Charles Seymour, for all the butter and sugar he could get? The answer, as it turns out, is certainly no. What Venizelos longed for, when not at home, was proper Cretan food. I can and will now tell you exactly what foods he wanted, from his own letters and the memoirs of people who knew him.

International Diplomacy and the Cretan Diet

His own fruit trees were what came into his mind in spring 1936, far away in exile, during the last two weeks of his life, when he wrote to remind a school friend to pick the plums and apricots from his garden: he didn't think he was going to be back in time. In purely chronological terms that is the last link between Venizelos and food that a biographer can make. The first link, I believe, is 39 years earlier, in spring 1897, when he was among the anti-Ottoman rebels on the Akrotiri peninsula, and a friend presented him with a hare. He knew just how to deal with it, and it went down well.

Between those two dates, and when Venizelos was living in Athens and Paris, the evidence is unequivocal. He longed for Cretan cheeses: Cretan *mizithra* and fresh young *athotyro* oozing butter. When it came to bread, he wanted it black, *mavro*; but he liked other Cretan breads and cakes, *koulourakia*, *bourekakia*, *kaltsounia* (little cheese tarts made with *mizithra*). He dreamed about the fruits of the island: the grapes, figs, and Cretan oranges. He dreamed of a Cretan vegetarian feast. *Sparangia*, asparagus, and not from his garden but wild Cretan asparagus (*Asparagus stipularis* and other species): the way he liked it prepared is (according to my Greek authority) 'ala vinegret, like the humblest Cretan peasant,' though one suspects the humble peasant did not call it *ala vinegret*. There was another asparagus-like delicacy, *avronies*, tender shoots of black bryony, *Tamus communis*, gathered from the fields; there would be *vrouves*, wild mustard greens, *Sinapis alba*, from the mountains. What was needed in Athens to make Venizelos's face light up (so his private secretary, Stefanos Stefanou, reported) was a basket of *chortarakia*, wild greens from the Big Island, and he would write a few words of thanks on a visiting card to any caller from Crete who brought him such a welcome gift. In that basket there might be *radikia*, dandelion leaves according to some (but that is rather *agrioradiki* or *pikralida*); wild chicory leaves, *Cichorium intybus*, according to others more authoritative. There might be *askolimbri*, golden thistle, *Scolymus hispanicus* and *Scolymus maculatus*. There might be the *avronies* and *vrouves* already mentioned. The ideal fate for the contents of that basket was to be turned into *vrastes salates*, boiled or blanched salads, to be eaten with olive oil, and not just any olive oil. Venizelos's palate demanded the heavy, biting, unfiltered oil of Crete, and he knew all about harvesting, preparing, and pressing it: his first wife (who died in childbirth in 1894 when he was still a Cretan lawyer) had been the daughter of an oil-merchant. He could talk for hours about olives and about that difficult issue that has worried olive-growers ever since ancient Greek times: how to get a good clean harvest without diminishing next year's crop. He liked table olives as well, Cretan olives, and particularly the tiny ones, the *elitses* that are called *koroneikes* in Crete; he liked *throumbes*, the windfalls that Cretans call *ambadiotikes* or *chamades*.

However much Venizelos liked Cretan salads with pungent olive oil, the evidence suggests that the millionairess Helena never learned to like them. Seven years after they married – he was Greek prime minister once more – she went with him to Rome in 1928 to sign the pact of friendship with Mussolini, and she describes a scene at dinner:

Mussolini, qui se redressait de toute sa petite taille, faisait ostentation de simplicité. Pendant un déjeuner il réclama à haute voix de l'huile pour la salade, disant: 'Je suis un paysan.'

If Venizelos spoke up to agree with him, Helena would not even have thought of mentioning the fact.

Bibliography
Apostolakis, S.A. *Laografika meletimata gia ton Elefth. K. Venizelo.* Chania: the author, 1995.
Lord Beaverbrook, *Men and power, 1917-1918.* London: Hutchinson, 1956.
Bonsal, Stephen. *Suitors and suppliants: The little nations at Versailles.* New York: Prentice-Hall, 1946.
Dalby, Andrew. *Eleftherios Venizelos: Greece (Makers of the modern world).* London: Haus, 2009.
Lambraki, Myrsini. *Ta chorta.* Chania: Trochalia, 1997.
Malcolm, Ian. *Lord Balfour: A Memory.* London: Macmillan, 1930.
Mee, Charles L. *The End of Order: Versailles 1919.* London: Secker & Warburg, 1981.
Nicolson, Harold. *Peacemaking 1919.* London: Constable, 1933.
Polunin, Oleg. *Flowers of Greece and the Balkans.* Oxford: Oxford University Press, 1980.
Seymour, Charles. *Letters from the Paris Peace Conference,* ed. Harold B. Whiteman. New Haven: Yale University Press, 1965.
Shotwell, James T. *At the Paris Peace Conference.* New York: Macmillan, 1937.
Steffens, Lincoln. *Autobiography.* New York: Harcourt, Brace, 1931.
Stevenson, Frances. *Lloyd George: A Diary,* ed. A.J.P. Taylor. London: Harper & Row, 1971.
Vardavas, C.I., D. Majchrzak, K.-H. Wagner, I. Elmadfa, and A. Kafatos, 'The antioxidant and phylloquinone content of wildly grown greens in Crete' in *Food chemistry* vol. 99 no. 4 (2006) pp. 813–821.
Lord Vansittart. *The Mist Procession.* London: Hutchinson, 1958.
Veniselos, Hélène. *A l'ombre de Veniselos.* Paris: Genin, 1955.
Zohary, Daniel and Maria Hopf. *Domestication of Plants in the Old World* 3rd ed. Oxford: Oxford University Press, 2000 (p. 139).

The Carrot Purple

Joel S. Denker

The orange carrot, now so familiar, was once a novelty. In fact, this young upstart was first cultivated a little more than 400 years ago. Until then, the purple variety was supreme. Although we consider the carrot immutable, it has been continually reinvented though the ages.

The weedy wild carrot is the oldest form of the plant. Scientists have uncovered wild carrot seeds at neolithic campsites in southern Germany and Switzerland dating back to 2500 BC. Unlike other members of its family—like dill, anise, coriander, and parsley, with their characteristic umbels (clusters of flowers) and aromatic seeds and roots—the carrot was not quickly cultivated. Its sisters, mostly grown in the Mediterranean and used as aromatics and medicines, seem to have been more highly valued.[1]

The wild carrot, which has a tiny, thin root, is the forebear of today's robust, fleshy vegetable. The plant, which sprouts prolifically in the fields and roadsides, came to be known as Queen Anne's lace. Since it is avidly chewed by animals, it is also known as 'cow's currency.' Wild carrot rapidly goes to seed; to protect its bounty of seeds, the plant's umbel closes, forming a 'bird's nest,' giving it another popular nickname.[2]

Its pretty white flowers and striking umbels, sometimes with more than 12,000 petals, make it especially attractive. In its center, a purple flower often springs up. The royal shade was said to have come from a single drop of blood that fell from Queen Anne's finger when she pricked it sewing lace.

The wild carrot was an unappealing food because of its bitterness. The carrot's acrid taste did win it favor as a medicine. The plant, whose botanical name, *daucus*, derived from the Greek word for burn, was a popular diuretic and stomach-soother. The seventeenth-century English herbalist Nicholas Culpepper praised its benefits to the body: 'Wild carrots belong to Mercury, and expel wind and remove stitches in the side, promote the flow of urine and women's courses, and break and expel the stone.... it helpeth conception.'[3] The flower was reputed to cure epilepsy.

The carrot has been long regarded as an aphrodisiac. The Greek name for the wild root was *philtron*, the word for loving. Many years later, the carrot retained its erotic mystique. Men in 1870s Teheran, food-writer Jane Grigson reports, swore that carrots stewed in sugar enhanced their potency.[4]

To noble ladies, the wild carrot also provided an alluring adornment. During the seventeenth-century reign of England's King James I, women of the court decorated their hair, their hats, and their dresses with its feathery fronds.

How did this primitive plant come to be domesticated? An expedition of Russian agronomists found a wide variety of carrots, both wild and cultivated, many a vivid purple, growing in Afghanistan in the 1920s. The domesticated carrot, they concluded, was first grown in the mountainous terrain, where the ridges of the Hindu Kush mountains and the Himalayas meet, in the tenth century AD.[5]

Farmers of the region, it is surmised, were unhappy with the scrawny, fork-rooted wild carrot. Through a process of selection, they were able to cultivate a tastier vegetable with thicker, smoother roots that was also more easily harvested.[6]

Purple carrots and yellow carrots, a likely mutant of the former, were introduced to the wide reaches of the Islamic empire by travelers and merchants. From Afghanistan, carrots traveled to Iran, where they were reported in the tenth century. These 'Eastern carrots' cropped up in Syria a hundred years later.[7] From North Africa, the vegetables reached Spain, where Ibn Al Awam, an Arab agriculturalist in the Islamic kingdom of Andalusia, observed them. The purple carrot, he found, was 'more succulent' than the yellow. He also highlighted the carrot's other desirable attributes. It was a 'diuretic' plant that 'increases the sexual appetites' and 'delights the heart.'[8]

Despite Al Awam's misgivings, the yellow carrot figured prominently in medieval Arabic cuisine. It was the foundation of *dinariyya*, a dish in which the root was sliced in disks that resembled the gold coin, the dinar.[9]

Purple and yellow carrots surfaced in China during the 1300s. The import likely came from Central Asia because it was dubbed the 'Iranian turnip.'[10]

The peregrinating carrot left the Mediterranean and took root in Holland, France, and Germany during the 1300s. A century later, Flemish exiles fleeing persecution in Holland transplanted the vegetable to England.

Initially, the carrot was cloaked in mystery. Many did not know what it looked like. 'Carrots are red roots sold by the handful in the market,' a Parisian man instructed his new wife in the fourteenth century.[11] (What we today describe as 'purple' was typically termed 'red'.)

After an early embrace, Europeans grew dissatisfied with the purple carrot. The issue was not flavor, because the pigment had no appreciable effect on taste. The purple's disadvantage was that its color seeped into sauces, soups, and stews, turning them a brownish purple.[12] The color also leaked on to the cook's hands. The yellow variety soon supplanted it.

But the reign of the yellow carrot was short-lived and an orange root soon took center stage. Otto Banga, a Dutch agronomist, documented the change. During the 1950s, Banga examined still-life paintings in the Louvre and other museums and detected an intriguing pattern. By the seventeenth century, orange carrots became more prominent in Dutch painting. He noted, for example, a bunch of long, pale orange-yellow carrots in a kitchen scene drawn by P.C. van Rijck in the early 1600s. During the course of the seventeenth century, Banga found a deeper orange shade in the carrots portrayed.[13]

The paintings demonstrated, Banga argued, that the Dutch had bred an orange carrot, selecting plants that had grown up in the population of yellow carrots and improving on them. 'Unlearned vegetable growers,' he suggested, were the pioneering cultivators.[14] The mostly female farmers in time produced four orange varieties, from which all similarly colored modern types descend.

Carrots were no longer simply carrots any more. Systematic breeding and classification of carrots originated during the seventeenth century. Carrots were categorized by the size, shape, and length of their roots, factors that increased their yield.

No-one has definitively solved the puzzle of the orange carrot. We know that it achieved Western supremacy but we don't know exactly why. The color took hold in Holland, it is argued, because of Dutch nationalism. This carrot, it is said, honored William of Orange and his House, because the orange variety was developed during his reign.[15] It was also excellent nourishment for farm animals. The Dutch attributed the creamy yellow butter made from their Holsteins' milk to a diet of orange carrots.[16]

But what explains the orange carrot's wider appeal? Philipp Simon's common-sense speculation may be the most convincing: 'The orange must have tasted pretty good or they might have gotten rid of it.' [17]

The versatile carrot became a staple in the food of many cultures, each of which put its own unique stamp on it. Carrots lent themselves to stews and soups but also to sweeter treatments. Dishes were also conceived that capitalized on the flavor of what many experts dub the second sweetest vegetable. In eighteenth-century Europe, food-writer Jane Grigson observes, this was not unusual: 'When new vegetables came in, they were viewed without savoury prejudice.'[18]

The English made puddings enriched with carrots. Hannah Glasse, the famous eighteenth-century English food writer, offered a recipe for carrot custard in puff pastry. The sweet was made with lavish amounts of cream and eggs and perfumed with nutmeg and orange-blossom water.[19]

Polish and Russian Jews celebrated New Year with *tzimmes*, a carrot stew. To the Jews, the carrot, one of the few sweet vegetables in their homelands, was auspicious. Fortuitously, their word for carrot, *mehren,* sounded like another Yiddish word, *mehrn*, which meant 'increase or multiply.' The Jews were delighted by the coincidence. To them, a feast of carrots promised prosperity and good fortune.[20]

The oldest carrot heartlands also contributed ingenious confections. Both the Afghans and the Persians have conjured up a carrot jam redolent of cardamom and flavored with pistachios and almonds.[21] The Moroccans play off carrots against the sweetish tang of oranges in a classic salad.

The modern carrot continues to evolve. Feeling the pressure to develop and market new products, growers have ceaselessly transformed the vegetable. In the 1950s, carrots, which were first sold in bulk, began to be sold in the US in cellophane bags. The 'cello carrot' required no change in the carrot's appearance. 'It had to look like a carrot and that was enough,' geneticist Dr Philipp Simon points out.[22] The latest marketing innovation, 'baby carrots,' demanded a reconfigured product and a change in emphasis on the kind of carrot grown.

Mike Yurosek, a California farmer, revolutionized the industry by dreaming up 'baby carrots.' Tired of throwing out 400 tons of imperfect carrots a day at his Bakersfield packing plant, he pondered making more efficient and profitable use of his crop. Freezing-plants bought his carrots and cut them into cubes, coins, and other shapes. 'If they can do that, why can't we, and pack 'em fresh,' he calculated.[23]

After first cutting them by hand in his own kitchen, he bought an industrial green-bean cutter from a failing frozen-food company. The machine turned out two-inch pieces of the root. By 1989, Yurosek had built a mechanized operation to turn out his product.

Marketed as 'baby carrots,' the miniatures are not young vegetables at all. 'They're grown-up carrots cut up into two-inch sections, pumped through water-filled pipes into whirling cement-mixer-size peelers and whittled down to the niblets Americans know, love, and scarf down by the bagful,' as journalist Elizabeth Weise observed.[24]

The 'fresh cut' segment of the carrot industry, which includes baby carrots and items like carrot chips, carrot sticks, and shredded carrots, is the fastest growing and most profitable. Baby carrots, the business's 'sweetheart,' as scientist Simon calls them, have spurred growers to concentrate on producing a long, thin, narrow vegetable.[25] A shorter, fatter root is less suitable for processing the high-value product. 'Prior to baby carrots, the ideal length for a carrot was somewhere between six and seven inches,' Simon notes.

An eight-inch carrot, a 'three cut,' can produce three two-inch babies. Further lengthening the carrot is a boon to the farmer. 'You make it a four-cut, and you've got a 33 per cent yield increase,' Simon says.[26]

Because fresh carrots must be mass-produced in elaborate, large-scale facilities, the industry is now the preserve of large corporations with deep pockets. Two firms, Boathouse and Grimway, control the market.[27]

Does the Western carrot of choice have to be orange? Perhaps there is a market for different shades of carrot? Dr Simon, a University of Wisconsin horticulture professor and geneticist, has been grappling with these questions. Simon, who directs the US Department of Agriculture's vegetable breeding program at the University of Wisconsin-Madison, is America's leading scientific breeder of carrots.

Simon, who studied potatoes as a graduate student, stumbled into carrot research because of his interest in genetics. The wide spectrum of colors among carrots piqued his curiosity. What genetic factors, he wondered, explained this phenomenon? 'I got interested because it was unusual,' he said. He jumped into his investigation before he realized that colors were of 'nutritional significance.'[28]

In an early experiment, Simon sharply increased beta carotene, the pigment in the orange carrot. Carotene is also used by the body to manufacture Vitamin A. The deeper orange root he produced contained four times as much of the coloring as the standard carrot. Manipulating the vegetable's hue, he was learning, conferred significant health benefits. 'About 30 per cent of the vitamin A we consume comes from carrots,' Simon told a reporter in 1995. 'It used to be 14 per cent about 25 years ago, but it has increased due to the higher beta carotene levels in carrots today.'[29]

Like beta carotene, the pigments responsible for other colored carrots also guard the body just as they shield plant cells during photosynthesis. Lycopene, the coloring in red carrots, protects against heart disease and may keep prostate cancer at bay. The xanthophylls in yellow carrots help insure healthy eyes. The anthocyanins, which make carrots purple, are strong antioxidants.[30]

The strange-colored vegetables, Simon recognized, had to be modified before they would be accepted. The purple carrot his laboratory acquired from Turkey, for example, had several handicaps. Since it was not as tasty as the orange, Simon crossed the two to make a more flavorful variety. The hybrid had another advantage. It resisted disease to which the purple was victim. 'We were fascinated by dark purple carrots from Turkey, but back in Wisconsin, they literally melted in the face of sclerosia, a pathogen that attacks many carrots but not orange ones,' Simon observed. 'We didn't even know this disease was still around.'[31]

In some cultures, the purple carrot is no oddity. Şalgam, one of Turkey's most popular refreshments, is a cool summer drink that relies on the root vegetable for its flavor. To make it, pickled carrots and turnips are fermented in barrels. Served in large glasses with a side dish of pickled carrots, it complements spicy kebab dishes. Şalgam also helps settle the stomach, its fans say, and softens the effects of raki, the intoxicating Turkish beverage.[32]

Dr Simon is still struggling to break down American resistance to multi-colored carrots. A purple carrot, however flavorful, makes consumers nervous, even though the shade has no effect on the taste. Before sampling his vegetables, Simon found, people would ask 'are they really carrots?' or 'are they safe to eat?'[33] 'We've become married to the colors we associate with particular foods,' he learned. 'We eat with our eyes, to some extent.'[34]

Meanwhile, Ersu, a Turkish company has begun bottling black carrot juice, extracted from what we would call purple carrots, for export to the US and other countries. The firm has great expectations for Black Miracle drink, one of its 'healthy, functional, and organic products.' [35] Perhaps there is still hope that the efforts of Dr Simon, who has two cases of purple carrot juice in his office, will not be in vain.

Acknowledgments

I am indebted to Dr. Philipp Simon, whom I interviewed extensively, for many of the insights on carrots expressed in this paper. Conversations with long-time friend and student of botany, Peter Adams, were most helpful. I appreciate the support of Peter Wolff, the editor of *The InTowner*, for encouraging my writings in food history.

Notes

1. Philipp W. Simon, 'Domestication, Historical Development, and Modern Breeding of Carrot,' pp. 171–72.
2. Among the most valuable sources on the wild carrot were Claire Shaver Haughton, pp. 313–316, and Barbara Perry Lawton, pp. 70–71 and pp. 110–112.
3. World Carrot Museum.
4. Jane Grigson, p. 161.
5. Otto Banga, *Main types of the Western Carotene Carrot and their origin*, pp. 13–17, and Philipp W. Simon, 'Domestication, Historical Development, and Modern Breeding of Carrot,' p. 166.
6. Author's interview with Simon, June 24, 2008.
7. Otto Banga, *Main types of the Western Carotene Carrot and their origin*, p. 19.
8. J.-J. Clément-Mullet, p. 178.
9. Personal correspondence with Charles Perry, August 29, 2008.
10. Berthold Laufer, p. 451.
11. Jane Grigson, p. 161.
12. Otto Banga, 'Carrot,' p. 292.
13. Otto Banga, *Main types of the Western Carotene Carrot and their origin*, pp. 21–25.
14. Ibid, p. 33.
15. 'Carrots, a Little History.'
16. Vegetarians in Paradise.
17. Author interview with Simon, June 11, 2008.
18. Jane Grigson.
19. Betty Fussell, p. 319.
20. Joan Nathan; Patti Shostek, p. 206.
21. Helen Saberi, p. 265; Margaret Shaida, p. 227.
22. Elizabeth Weise.
23. Ibid.
24. Ibid.
25. Author interview with Simon, June 11, 2008.
26. Elizabeth Weise.
27. Gary Lucier and Biing-Hwan Lin.
28. Author interview with Simon, June 11, 2008.
29. Karen Herzog, p. 1.
30. See Karen Herzog, Pam Nevar, and Jude Stewart for analyses of the role of carrot pigments.
31. Jude Stewart.
32. Burak Sansal.
33. Author interview with Simon, June 11, 2008.
34. Erin Peabody.
35. Ersu, 'New Star of Ersu: Blackish Miracle.'

Bibliography

Books

Behr, Edward. *The Artful Eater: A Gourmet Investigates the Ingredients of Great Food*. Peacham, VT: The Art of Eating, 2004.

Banga, Otto. 'Carrot.' In *Evolution of Crop Plants*, edited by N.W. Simmonds, 291–293. London: Longman, 1926.

———. *Main types of the Western Carotene Carrot and their origin*. Zwolle, The Netherlands: W. E. J. Tjeenk Willink, 1963.

Clément-Mullet, J.-J. *Le Livre de L'Agriculture*. Tunis: Les Editions Bouslama, 1977.
Davidson, Alan. *The Oxford Companion to Food*. New York: Oxford University Press, 1999.
Fussell, Betty. *I Hear America Cooking*. New York: Viking, 1986.
Greene, Bert. *Greene on Greens*. New York: Workman Publishing, 1984.
Grigson, Jane. *Jane Grigson's Vegetable Book*. Lincoln: University of Nebraska Press, 2007.
Haughton, Claire Shaver. *Green Immigrants: The Plants that Transformed America*. New York: Harcourt Brace, 1978.
Hyams, Edward. *Plants in the Service of Man: 10,000 Years of Domestication*. London: J.M. Dent & Sons, 1971.
Laufer, Berthold. *Sino-Iranica: Chinese Contributions to the History of Civilization in Ancient Iran*, Publication 201. Chicago: The Field Museum of Natural History, 1919.
Lawton, Barbara Perry. *Parsleys, Fennels, and Queen Anne's Lace: Herbs and Ornamentals from the Umbel Family*. Portland, Oregon: Timber Press, 2007.
Lovelock, Yann. *The Vegetable Book: An Unnatural History*. New York: St. Martin's Press, 1972.
McNamee, Gregory. *Moveable Feasts: The History, Science, and Lore of Food*. Lincoln: University of Nebraska Press, 2008.
Roberts, Jonathan. *The Origins of Fruit and Vegetables*. New York: Universe Publishing, 2001.
Roden, Claudia. *A New Book of Middle Eastern Foods*. New York: Penguin Books, 1985.
Root, Waverly. *Food: An Authoritative and Visual History and Dictionary of the Foods of the World*. New York: Smithmark Publishers, 1996.
Rubatzky, V.E., Quiros, C.F., and Simon, P.W. *Carrots and Related Umbelliferae*. Wallingford, England: CABI Publishing, 1999.
Saberi, Helen. *Afghan Food and Cookery*. New York: Hippocrene Books, 2000.
Shaida, Margaret. *The Legendary Cuisine of Persia*. Brooklyn, NY: Interlink Books, 2002.
Shostek, Patti. *A Lexicon of Jewish Cooking: A Collection of Folklore, Foodlore, History, Customs and Recipes*. Chicago: Contemporary Books, 1981.
Simon, Philipp W. 'Domestication, Historical Development, and Modern Breeding of Carrot.' In *Planting and Breeding Reviews Volume 19*, edited by Jules Janick, pp. 157–190. New York: John Wiley, 2000.
Simoons, Frederick J. *Food in China: A Cultural and Historical Inquiry*. Boca Raton, FL: CRC Press, 1991.
Sumner, Judith. *American Household Botany: A History of Useful Plants, 1620–1900*. Portland, OR: Timber Press, 2004.
Toussaint-Samat, Maguelonne. *History of Food*. Malden, Massachusetts: Blackwell Publishers, 1998.

Articles

Bonné, Jon. 'Convenient Carrot Charms Consumers: Peeled and ready-to-eat transforms an unheralded vegetable.' MSNBC, July 23, 2003. Available online at http://www.msnbc.msn.com/id/3072775/. Accessed June 10, 2008.
Brow Farm. 'Carrots.' Available online at http://www.browfarm.co.uk/carrots_about.htm. Accessed June 1, 2008.
'Carrot, Daucus carota, Origin and Archeology of Carrot, Modern Researches.' MDidea Exporting Division. Available online at
http://www.mdidea.com/products/new/new069paper.html. Accessed June 8, 2008.
'Carrots, a little History.' HungryMonster.com. Available online at http://www.hungrymonster.com/FoodFacts/Food_Facts.cfm?Phrase_vch=Carrots&fid=6091. Accessed November 10, 2007.
Ersu. 'New Star of Ersu: Blackish Miracle.' Available online at http://www.ersu-nar.com/indexen.php?goster=carrot&cond=bos. Accessed June 22, 2008.
Fiszer, Louise and Ferrary, Jeannette. 'Carrots.' Available online at http://www.sallys-place.com/food/columns/ferray_fiszer/carrots.htm. Accessed November 19, 2007.
Griffin, Kawanza L. 'Root Work: Carrots' nutrients can be judged by the color of their skin, studies find.'

Milwaukee Journal-Sentinel Online. Available online at http://www2.jsonline.com/alive/nutrition/oct01/carrots29102801.asp. Accessed June 18, 2008.

Herzog, Karen. 'A supercarrot, purple carrot, and other root causes: Why the familiar orange article is a happy accident.' *Milwaukee Journal-Sentinel*. Sunday, 29 October 1995, Lifestyle and Food, p. 1.

Kulu, Şule. 'Traditional Turkish beverages storm domestic market again.' *Sunday's Zaman*, 22 July 2007. Available online at http://www2.jsonline.com/alive/nutrition/oct01/carrots29102801.asp. Accessed June 11, 2008.

Lucier, Gary and Lin, Biing-Hwan. 'Factors Affecting Carrot Consumption in the United States.' United States Department of Agriculture, Economic Research Service. Available online at http://www.ers.usda.gov/publications/vgs/2007/03Mar/VGS31901/vgs31901.pdf. Accessed June 11, 2008.

McCall, Celeste. 'How Sweet They Are.' *The Washington Times*, Sunday, 12 January 1992, Part E, p. E1.

Nathan, Joan. 'Tsimmes worth the big fuss.' *Milwaukee Journal Sentinel*. Sunday, September 24, 2000, ENTREE, p. 01N.

Nevar, Pam. 'Pigment Power in Carrot Color.' College of Agricultural and Life Sciences, University of Wisconsin-Madison. Available online at http://www.cals.wisc.edu/media/news/02_00/carrot_pigment.html. Accessed May 7, 2008.

Peabody, Erin. 'Carrots with Character.' *Agricultural Research*, 1 November 2004.

Sansal, Burak. 'Traditional Turkish drinks.' AllaboutTurkey.com. Available online at http://www.allaboutturkey.com/icecekler.htm. Accessed June 11, 2008.

Simon, Phillip W. 'Carrot Facts.' U.S. Department of Agriculture, Agricultural Research Service, Vegetable Crops Unit, available online at http://www.ars.usda.gov/Research/docs.htm?docid=5231. Accessed May 27, 2008.

Stallsmith, Audrey. 'Queen Anne's Carrot.' Suite101.com. Available online at http://www.suite101.com/article.cfm/historical_plants/117382. Accessed May 26, 2008.

Stewart, Jude. 'Feast Your Eyes: Brilliantly colored, everyday vegetables hit the middle-market.' *STEP Inside Design*, 2006. Available online at http://judestewart.com/writing/stepvegetables.htm. Accessed June 10, 2008.

Turkish Daily News. 'Beverages that bring the scent of Anatolia.' December 1, 2007. Available online at http://www.turkishdailynews/com.tr/article.php?enewsid=90041. Accessed June 11, 2008.

Vegetarians in Paradise. 'And the 24 Carrot Award Goes to . . .' Available online at http://www.vegparadise.com/highestperch412.html. Accessed November 10, 2007.

Weise, Elizabeth. 'Digging the baby carrot.' *USA Today*. 11 August 2004, Lifestyle.

World Carrot Museum. 'History of the Carrot – Origins and Development.' Available online at http://www.carrotmuseum.co.uk/history.html. Accessed November 10, 2007.

———. 'Carrot History Part Two – From Medicine to Food, A.D. 200 to 1800.' Available online at http://www.carrotmuseum.co.uk/history2.html. Accessed November 10, 2007.

———. 'Carrot History Part Three – Evolution and Improvement, 1800 to date.' Available online at http://www.carrotmuseum.co.uk/history3.html. Accessed November 10, 2007.

Yaka Organics. 'Original Black Carrot with High Color Content from Turkey.' Available online at http://www.yakagroup.com/black_carrot_black_carrot.html. Accessed June 11, 2008.

Listening to Vegetables

Len Fisher and Nick Sorensen

The world of vegetable acoustics is unfamiliar to most people, but it covers:

> The screams, groans, and agonized poppings that root and leaf vegetables make as they dry out after harvesting.
>
> Acoustic testing of vegetables, from the laboratory to the supermarket.
>
> Vegetable music, including a recording of the Vienna Vegetable Orchestra (every instrument is made from a vegetable!), a demonstration of the carrot clarinet, and a presentation of 'Carrot Crunch' by contemporary musician Philip Corner.

Spontaneous Sounds

'I talk to the trees' goes the well-known song, but few people realize that the trees can talk back. Cornish artist Alex Metcalf has made their voices available to everyone via his 'tree listening installation' in the Royal Botanic Gardens at Kew.[1] Sensitive microphones attached to the trees send their amplified signals to headphones, allowing the listener to hear every snap, crackle and pop that the trees are making.

Those sounds reflect the stress that the tree is under. The hotter the day, the more frequently the tree cries out. The source of the sounds is cavitation – the breaking of liquid xylem columns under tension. The columns are being stretched by capillary forces in the leaves. The tension increases as the plant dries out, and when it reaches a critical value the column breaks with a tiny supersonic bang that the microphones can pick up.

The more frequently those sounds occur, the drier the plant has become. Horticulturalists are now using the sounds to monitor the condition of green house crops,[2] with the eventual plan (now being realized in some commercial greenhouses) of using the measurements to control watering regimes. The same principle is also being used to monitor the condition of field crops such as corn and to control the irrigation of these crops in a more efficient manner.[3]

Acoustic Testing

We don't have to rely on the spontaneous sounds that plants make. We can tap, hit, pummel, or otherwise disturb the plants, and listen to the sounds that they make in response. This approach is called 'acoustic testing,' and it can tell us a great deal about the internal condition of a fruit or vegetable without doing the damage that more invasive tests impart.

Listening to Vegetables

'Acoustic testing' is something that we often do without realizing it. When we bite into a fresh, crisp apple and enjoy the evocative sounds that go with that sharp, delicious taste, we are doing our own form of acoustic testing. When we tap a watermelon in a supermarket and listen for the sound that it makes, we are more obviously doing acoustic testing. A ripe, fresh melon should sound hollow and a little muffled, reflecting the high turgor in the cells and the very slight softening in the centre. [4]

In the scientist's hands, acoustic testing goes a good deal further. It has been used, for example, to measure microscopic losses of firmness (as little as 0.1%) caused by drying out of carrots and other hard vegetables [5] (and hence to determine just how fresh they are) and to follow the ripening and predict the shelf life of apples and tomatoes in the supermarket.[6] A simple search of Google Scholar using the keywords 'acoustic' AND 'optimum' AND 'ripening' reveals how widely these techniques are now being used, with nearly a thousand hits.

The techniques are all scientific variants of tapping on a watermelon. They subject the vegetable to a slight mechanical stress and listen to the sounds that it makes in response. The stress need not be too large – the vegetable equivalent of gentle questioning in a friendly *viva*, rather than interrogation in Guantanamo Bay-style. The vegetable will respond to such kind treatment by giving all manner of information about itself. Small cracking sounds will tell the experienced listener how well the cell walls can stand up to stress, and by inference how crunchy the vegetable will be when eaten (the stronger the cell wall, the louder the noise that it makes when it fractures under stress, and the crunchier the vegetable will be). Damped-out sounds mean a soft interior, and other changes to the sound pattern can mean a loss of interior juices. All of these can be assessed by eating the vegetable, but the point of acoustic testing is first to select the best vegetable to eat.

That's if you want to eat the vegetable in the first place. Vegetables also have other uses, especially in the arts. Filmmakers have been known to line an empty room with plastic sheets and pulverize a variety of fresh produce with hammers and mallets to capture the sounds of breaking bones, splattering blood, and tearing flesh.[7] Other artistic uses of vegetables have included French artist Ben Vautier's installation of decomposing fruit and vegetables under glass and the game of vegetable chess played during the 2007 'Flux-Olympiad' at the Tate Modern.

Surely the most unusual use of vegetables, though, has been as musical instruments. The use of vegetables and vegetable matter as source material for instrument makers and musicians has a long history; a history that commences with the earliest examples of musical instruments continues right up to, and includes, contemporary performance.

The Use of Vegetables in the Construction of Musical Instruments

How do vegetable instruments work? What is their secret? In truth, it is fairly simple. The main criterion is that the chosen vegetable must be able to spring back into shape quickly after deformation. In other words, it needs to be *elastic*.

Listening to Vegetables

All vegetables (in fact, all materials) have a combination of *elastic* and *viscous* properties – in other words, they are *viscoelastic*. *Elastic* means the ability to rebound to the original shape after deformation followed by removal of the deforming force. *Viscous* means that the material will flow when it is deformed, and so will not return to its original shape when the deforming force is removed. Contrasting examples from the world of edible plants are a stick of celery and a raspberry. If you put your celery on top of your raspberries in your shopping bag, the raspberries will become squashed and remain squashed. If you put the same weight of raspberries on top of the celery, the celery sticks will bend, but spring back into shape when the raspberries are lifted off.

To turn a vegetable into a musical instrument, the emphasis needs to be on elasticity, because elasticity means that no energy is lost, while viscous deformation soaks up energy and dampens the sound (as it does in an over-ripe watermelon). The material also needs to be reasonably hard (i.e. to deform only slightly under stress) so that it will keep its shape, and hence its tone.

The very earliest musical instrument was probably a hollowed out log; hit with sticks, it became a prototype drum. Following this we have many other examples of musical instruments that have been made entirely of vegetables and/or vegetable matter, or have used vegetables as part of their construction. A survey of these instruments highlights the use of two vegetables in particular: the gourd and bamboo, both of which are hard and elastic. Examples of the way these materials have been used can be found throughout the world.

Gourds have been used extensively in the creation of musical instruments. They make ideal rattles. In Africa gourds and calabashes have been used as rattles, sometimes covered in beads. Filled with dried seeds these shakers have a magical potency. In India women lull children to sleep with a gourd rattle although they are careful to use one that does not have secret magical objects placed inside it. The west African water drum comprises a pail-sized receptacle that contains water on which is floated a smaller bowl (usually a gourd). The smaller bowl is placed upside down and hit with a beater.

The central African xylophone uses the gourd to amplify the sounds of wooden bars. There are many examples of stringed instruments that use gourds for the body: the *kemancha*, the traditional fiddle of the Middle East from Iran; the *kora* from west Africa, a cross between a harp and a lute that is played in Gambia and Mali; the sitar from India. Ocarinas have been found made from dried fruit shells.

Bamboo is also used extensively and is an ideal material for creating flutes or percussion instruments, especially scrapers. The *angklung* is an Indonesian instrument made of swinging bamboo tubes. The *guiro*, a scraped instrument from South America, is made by making transverse notches along the surface of a section of bamboo. In south-east Asia and the Pacific Islands bamboo is used to make flutes; the Polynesian nose flute. Likewise we find panpipes in western Bolivia, the *shakuhachi*, an end-blown flute in Japan and the *xiao* in China. All made from bamboo. The use of bamboo is not confined to traditional or ethnic musical instruments. Yamaha currently produce a steel-strung guitar made from bamboo – an easily replenishable resource in place of expensive and rare hardwoods.

Whilst vegetables are most suited to create shakers, flutes and provide amplification chambers there are examples of vegetable matter used to create string instruments. The African corn-stalk zither uses corn stalks, bound together with plaited grass, as strings. These are stretched over a narrow bridge and played as a dulcimer.

The Vienna Vegetable Orchestra

In all the previous examples vegetables have been dried and prepared in some way in order to become suitable materials for the construction of musical instruments. The Vienna Vegetable Orchestra (founded in 1998 and based in Vienna),[8] however, create their instruments from fresh vegetables.

The hardness and elasticity in this case comes from choosing fresh vegetables which have very high turgor (i.e. internal cell pressure), and also from selecting vegetables that have very stiff and rigid cell walls. With carrot flutes, pumpkin basses, leek violins, leek-zucchini-vibrators, cucumberophones, and celery bongos, the orchestra creates its own extraordinary and unique sound world.

The orchestra plays concerts all over the world, and has no musical boundaries. Its diverse musical styles include contemporary music, beat-oriented House tracks, experimental Electronic, Free Jazz, Noise, Dub and Clicks'n'Cuts. The musical scope of the ensemble expands consistently, and recently developed vegetable instruments and their inherent sounds often determine the direction of their music.

The members of the ensemble are all active in various artistic areas and have worked together on conceptualizing and carrying out their project to develop vegetable music. The intention is to create a sonorous experience that can be perceived with all senses. The instruments are made from scratch just one hour prior to each performance using the freshest vegetables available, then all ninety pounds of vegetables are cooked into a soup following the performance (but see note 9).

The review of a concert in Singapore in 2006 captures the flavour of this ensemble

> playing with food never had such interesting results…the concert was truly a sensory experience towards the end when the sharp scent of celery and onion filled the venue and juices from bruised vegetables stained the performers' attire.…But this orchestra revelled in sound and as one left the concert more attuned to the aural rhythms of everyday life, it seems their devotion had paid off.[10]

[The original talk at this point presented a demonstration of the manufacture and playing of a 'carrot clarinet,' where a large carrot was drilled out along its length, provided with finger holes, a bell made from a plastic kitchen funnel, and a saxophone mouthpiece, and played by N.S.]

'Carrot Crunch'

The Vegetable Orchestra emerged from the mid-twentieth-century Fluxus movement, which was (and is) an informal movement of avant-garde artists who base their work on wit and 'childlikeness,' sometimes displayed through 'happenings' that combined music, theatre, performance art and poetry (note 11). The movement was loosely organized at its beginning in the early 1960s by the American-resident, Lithuanian-born artist George Maciunas, and later included artists such as Yoko Ono. Fluxus encouraged a do-it-yourself aesthetic, valuing simplicity over complexity. They focussed on using materials that were at hand; frequently their work comprised a written description of actions that were to be performed. These 'scores' were usually brief and simple, sometimes framing everyday acts as artistic performances.

It is the influence and legacy of this group of artists that provide the final example of vegetable music. The origins of their approach to artistic creativity can be traced to many of the concepts explored by the composer John Cage in his experimental music of the 1950s. An example of this approach to music making is 'Carrot Crunch' written by the American composer Philip Corner in 1964. Philip Corner was born in 1933, and he studied with Henry Cowell and Olivier Messiaen, amongst others. He began his association with Fluxus in 1961. Many of his scores are open-ended in that some elements are specified but others are left partially or entirely to the discretion of the performers. Some of his work employs standard musical notation whereas others utilize graphic or text based scores. His music frequently explores unintentional sound and chance activities.

'Carrot Crunch,' sometimes referred to as 'Carrot Chew Performance,' is a text-based piece. The performer is instructed to eat a carrot and for the sounds to be amplified through the use of a microphone and loudspeakers (note 12 gives a non-vegetarian alternative). The piece commences when the first bite is made into the carrot, and finishes when the final piece of the carrot is swallowed. This piece is typical of the Fluxus movement; it is not concerned with making big or complex gestures but instead is artist-centred, focusing on brief and simple events.

[We were unable to locate a copy of a 'score' for Carrot Crunch, but N.S. produced a performance based on one that he witnessed by the group Apartment House, a group specializing in avant-garde and experimental music from around the world. It was part of a concert called In The Shadow of Black Mountain, presented at the Arnolfini Gallery, Bristol on 14 January 2006.]

Notes

1. Alex Metcalf 'Tree Listening.' http://www.alexmetcalf.co.uk/AlexMetcalf/Tree_Listening1.html
2. David L. Ehreta, Anthony Laub, Shabtai Bittmana, Wei Lina and Tim Shelford. 'Automated monitoring of greenhouse crops.' *Agronomie* 21 (2001) 403-414.
3. Melvin T. Tyree, Edwin L. Fiscus, S. D. Wullschleger and M. A. Dixon. 'Detection of Xylem Cavitation in Corn under Field Conditions,' *Plant Physiol.* 82 (1986), 597-599.
4. Anon 'Taste,' http://www.taste.co.nz/Home/Howto/InSeason/tabid/288/ArticleID/1500/Default.aspx
5. József Felföldi & András Fekete. 'Detection of Small-Scale Mechanical Changes by Acoustic Measuring System,' Paper 036097, *American Society of Agricultural and Biological Engineers Annual Meeting* (2003).
6. Artur Zdunek & Zbigniew Ranachowski. 'Acoustic Emission in Puncture Test of Apples During Shelf-Life.' *Electronic Journal of Polish Agricultural Universities* Vol. 9, Issue 4 (2006).
7. Dark Sector Blog 'Vegetable Abuse.' http://darksectorgame.blogspot.com/2007/03/vegetable-abuse.html
8. Vienna Vegetable Orchestra. http://www.gemueseorchester.org/
9. Unless health and safety regulations dictate otherwise. When the orchestra performed at the Huddersfield Music Festival in November 2007, the soup-making was banned – even without the vegetable parts that the orchestra members had blown on (Source: 'Britain awaits the Vienna Vegetable Orchestra.' *Daily Telegraph*. 17 November 2007.)
10. Review, *The Straits Times*, Singapore, 10 June 2006.
11. An excellent, if unintentional, example of food-related *fluxus* is surely the annual festival of the radishes in Oaxaca, Mexico – a competition to see who can carve radishes into the most artistic shapes, with the results displayed in the town square while the proud artists stand by with water sprays to keep their creations fresh in the heat.
12. This is not the only piece of music that is based on the activity of eating. A non-vegetarian option was created by the performance artist Mark McGowan in May 2007 as a protest against the Duke of Edinburgh's involvement in killing foxes with the aid of dogs. McGowan's response in a radio broadcast was to eat a (cooked) corgi that had died on a breeding farm. The corgi was minced, mixed with herbs, apple, bread and onion, and served in pita bread with salad.

Vegetables as a Symbol in Design and Art

Anna Marie Fisker & Tenna Doktor Olsen

Photograph: Anna Marie Fisker.

Why do we human beings harbour desires that outweigh economic considerations by far? Why do we sometimes throw caution and common sense to the wind? Because we fall in love is the simple answer. Because we are fascinated by objects, their appearance, their tactile qualities, their shape, their function, and not least their essence.

Similarly, our historical drive toward the unknown, the exotic, toward possessing the object(s) of our desire has manifested itself in our love for porcelain. Historically, this is evident in, for example, European Renaissance banquets where kings and nobles decorated the walls and dining tables of their palaces with staggering displays of tableware; partly in order to exhibit their prized wealth, but also to dazzle and seduce viewers and dining guests with fabulous designs and unique craftsmanship.

The ornamentation of the room and the great displays of treasures move the meal and particularly the porcelain from their regular context into a cultural sphere where the meaning of the design of the porcelain takes on an importance equal to or greater than the practical function of the individual plate during the meal itself. Consequently, for special occasions, the porcelain provided the framework for the party and almost constituted the architectural platform on which the meals in the host's honour took place. It thus functioned as a political and social display of power or a symbol of wealth in addition to serving as practical tableware.

Porcelain is thus a coveted treasure. This unique material, these transparent and translucent objects, these forms made of the dreams that any European king would chase in his craving for status and power. Porcelain tableware, 'white gold', became objects that any prince worth his salt *had* to incorporate into his art collection.

Porcelain was originally invented and developed in China in the tenth century AD. The gem-like, white, semi-transparent material arrived in Europe in the fifteenth century with traders on the Silk Route to the West and achieved its lofty status at the European courts. During the next 200 years, however, the Europeans had to import porcelain from the East as they were unable to produce it and, because they wanted to know the secret behind these treasures, European princes spent vast amounts of money attempting to find the right chemical formula. In addition to this fascination, Europe witnessed countless attempts at figuring out the secrets of porcelain by people who wanted to be able to produce and manufacture it themselves. In Denmark, these efforts culminated with the establishment of *Den Kongelige Priviligerede Danske Porcelænsfabrik* ('Royal Copenhagen') on 1 May 1775.

In approximately 1710, the Bohemian alchemist J. F. Böttger succeeded in producing the authentic, white, hard porcelain that was subsequently produced at the factory in Meissen, Germany. Danish experiments to produce porcelain did not begin until the 1760s, but this was soft porcelain, without kaolinite, inspired by the French tradition, and it was a Frenchman, the sculptor Louis-François Fournier, who headed the Danish experiments. Fournier had much experience from his work at the porcelain factory in Sèvres outside Paris, but in spite of its quality, the porcelain he produced was not very durable and too expensive, so the production stopped in 1765.

In 1775, the company that would develop into Den Kongelige Porcelænsfabrik, now Royal Copenhagen, began its production of hard porcelain in the German mould. The history of the *Flora Danica* porcelain series begins in 1790 when King Christian VII ordered the first *Flora Danica* set from Den Kongelige Danske Porcelænsfabrik. The story goes that *Flora Danica* was originally intended as a gift from the Danish king to the Russian Empress Catharine II as Christian VII was worried by the possibility of war with Sweden and Norway and therefore wanted to strengthen his alliance with Russia. Thus the *Flora Danica* tableware was to be a diplomatic gift to the Russian Empress. It was a time when gifts among royals were supposed to be literally royal. The set was therefore meant to be a symbol of the very finest in Danish handcrafted products and, as would later become apparent, botany.

The set of tableware was ordered in 1790, but as the Empress died in 1796, before it was finished, the order was transferred to the Danish Crown Prince himself. The tableware

Vegetables as a Symbol in Design and Art

was delivered in 1802, and these original pieces remained the property of the Danish royal family where it was used only at special royal banquets, birthdays, and weddings.

What makes the *Flora Danica* tableware special is first and foremost the ornamentation of the individual pieces, which consist of no less than 1260 different motifs of Danish plants: herbs, flowers, and vegetables that are accurately depicted on the many pieces of the set: 1,802 pieces in total, 100 place settings. Technically and formally, the *Flora Danica* tableware was modelled on a pre-existing porcelain mould, the so-called Pearl Model, which was created by an anonymous designer and went into production for the first time in 1783.

The Pearl Model is smooth with a discreet relief around the edges. It represented a new style for the factory, a break away from wavy, curving Rococo forms. The model has a stylistic tightness that heralds neo-classicism. The decorative scheme echoes the classicism of the base design, with simple, geometric elements accentuated by gilding and a rose-pink blush. One might speculate that the Pearl Model might have been chosen because its form was very well suited for large decorations, and many of the depicted plants filled the entire surface.

It is not only the technical finesse and stylistic innovation, however, that make the set unique, but especially the accurate realism of the plant motifs. Small pieces such as knife handles, ash trays, etc., often have decorations that portray single buds or leaves of plants that are depicted in their entirety on the larger tableware; an expression of the designer's (and contemporary science's) taxonomic enthusiasm. The decoration of the tableware was in keeping with the spirit of the Enlightenment, and was chosen according to natural-scientific criteria, rather than as an expression of aesthetic judgement.

To the philosopher Immanuel Kant, the aim of enlightenment was 'mankind's release from self-inflicted lack of authority.' Reason, and the critical mindset that follows, alone must decide the accuracy or fallacy of every realization as well as the norms of ethical, political, and social acts.

The model for this future was created by Jean-Jacques Rousseau, among others. His primary idea was that mankind is inherently good but is perverted by culture, and he defined a 'back-to-nature' concept. Rousseau's perception of nature and culture was absolute. He distinguished between false nature and true nature; truth being something that stems from that which is natural to man. He argued that man and nature both must be shaped, but it must take place in accordance to natural drives. The key word is 'simplicity' in the sense that the simple is natural. He thus believed in emotion over reason. *Flora Danica* was created from this line of thought in terms of decorative value.

Flower painting on porcelain was not unusual at the end of the eighteenth century, but the flowers on the porcelain usually had to conform to aesthetic criteria. For the production of the *Flora Danica* tableware, however, 'scientific' illustrations were chosen: exact copies of plants from the *Flora Danica* botanical book.

The *Flora Danica* tableware was furnished with exact depictions of wild plants engraved as copperplates.[1] The models for these engravings were made by two illustrators from Nuremberg, Michael and Martin Rössler. These correct and accurate, almost pedantic, renderings were necessary in order to be able to classify them. In the great *Flora Danica* book itself, the illustrations had to be as accurate as possible, and the same goal was adopted when the motifs were conveyed to the porcelain.

The collected tableware was meant to appear as an illustrated Danish flora on the dining table. Therefore, all types of plants and vegetables were included, not just decorative flowers but also the less obviously decorative grass, mosses, and fungi – all plants were of equal importance from a natural scientific perspective.

In some instances, particularly large plants have been illustrated in two parts, which results in a highly unconventional decoration. This preference for botanical accuracy over aesthetic beauty is noteworthy, and the parameters for the design might outline the potential possibilities.

The idea of decorating a lavish and exclusive set of tableware with leaves, flowers, stalks and roots in a botanically correct rendition stemmed from a botanical work that was printed in Copenhagen from 1761 under the direction of the German-born botanist Georg Christian Oeder, royal professor of botany. *Flora Danica* is one of the world's biggest flora – a true child of the Enlightenment – it was 123 years in the making. Immediately on his appointment in 1753, Oeder began planning an illustrated folio, as well as laying out a new botanic garden near Amaliegade in Copenhagen. Both pprojects were undertaken with a view to better assess the flora of a realm that at that time included Denmark itself, the duchies of Schleswig and Holstein, Oldenburg and Delmenhurst in Saxony, as well as Norway and its dependencies in the North Atlantic: Iceland, the Faeroe Islands and Greenland.

Oeder's proposal was for folio depictions of every plant indigenous to the realm. The purpose was to spread knowledge of botany and thereby achieve greater knowledge of the useful and harmful properties of various plants, in medicine as well as gastronomy.[2]

During the Renaissance, when botany was reinvigorated by a news spirit of scientific enquiry, there were no rules as to how to describe and name plants. Thus, one plant had many names. The Swedish natural scientist Carl von Linné (1707–1778), known to the English as Carolus Linneaus, set out to give order to botanical classifications. In 1753, he published his famous work *Systema Naturae* in which he subdivides and classifies plants and animals according to new categories such as domain, realm, line, order, family, stock, and species. Linné's principles of classification also form the basis for the plates of *Flora Danica* and all subsequent plant classifications.

The *Flora Danica* tableware became important because the botanical correctness of the decorations of the dishes makes the set original. As mentioned, flower motifs were common in European porcelain from the 1770s, but composite bouquets or decorative flower spreads were more common than reproductions of entire plants with roots. Nevertheless, isolated parallels do exist in German and English porcelain sets with realistic plant renditions and with various birds and fruits.

Vegetables as a Symbol in Design and Art

In 1862, production of *Flora Danica* was recommenced on the occasion of Christian IX's daughter Alexandra's wedding to the Prince of Wales, later King Edward VII of Great Britain. Sixty place-settings were ordered. The design parameters had changed, though, as the book *Flora Danica* was still in progress – it was not finished until 1874. The choice this time was more aesthetic, and the most decorative and colourful flower decorations were selected for the set. Decorative grace was given higher priority than botanical correctness in the choice and renditions of plants. One characteristic of Princess Alexandra's set is that the Latin name of the plant is written on the back of the plates in beautiful calligraphy rather than the reference numbers that are found on the original set. The later porcelain was designed as an object of beauty rather than an informative work about Danish flora.

Until the end of the nineteenth century, the *Flora Danica* porcelain remained exclusive to the royal household, but the tableware was brought very much to the attention of the rest of the world when 75 pieces of the original set were exhibited in Paris in 1878.

Today, pieces of the fantastic *Flora Danica* tableware are still produced, and the individual pieces are now expensive collectors' items. One thousand five hundred and thirty pieces of the original set have been preserved. They are exhibited at Rosenborg Castle and Amalienborg Castle as well as at the Danish Museum of Art and Design and a number of other Danish museums.

As in all fairy tales and beautiful dreams, there must be room for something different, something that turns dreams into reality. Thus, the Danish designer and fashion artist Anette Meyer commenced her creation of a new and unique *Flora Danica* work. Meyer, who works with fashion and product design, created a beautiful paper dress in a pattern modelled on the principles of *Flora Danica*. The project is titled 'Icon Dressed' and shows the last 200 years of female fashion. Meyer emphasizes *Flora Danica* in 14 suits that have all been sewn in paper with the *Flora Danica* motif.

Flora Danica, in dress and on porcelain, depicts the history of particular food plants and the effect of these not only on the table, but also in the meal.

Notes

1. But how was it possible to use the round, curved or oval shapes on the porcelain in a coherent motif? Well, designers and painters obviously had to compromise the square illustrations from the books from time to time. Johann Cristoph Bayer, who had been called upon by Oeder as painter for the book *Flora Danica*, was chosen as flower painter. The *Flora Danica* tableware became Bayer's life work. In the original set, almost all the flowers were painted by Bayer, who was German and had been employed at Den Kongelige Porcelænsfabrik in 1776, himself. From 1790, he devoted himself entirely to the *Flora Danica* tableware, and it was not until the final phase before the completion of the set in 1802 supplemented by Nicolai Christian Faxøe who painted 158 pieces.
2. Oeder was dismissed from the order in the wake of the fall of Struensee in 1772, and the publication of the work was subsequently left to the zoologist O.F. Müller. Eight different botanists followed him in managing the publication of F*lora Danica*. The main part of the work – the part that is usually thought of when F*lora Danica* is mentioned – consists of 51 instalments plus 3 supplements with a total of 3,240 copper engraved plates. *Flora Danica* does not follow botanical systems. It can therefore be difficult to negotiate, but in 1887 the botanist Johann Lange published a 'Nomenclature, F*lora Danica…*' containing alphabetic, systematic and chronological overviews of the plants of *Flora Danica*.

References

Bencard, Mogens et al. (2005). *Dansk Porcelæn 1775–2000, design i 225 år*. Nyt Nordisk Forlag/Arnold Busck.

Dickson, Thomas. *Dansk Design*. Gyldendal, 1.ed., 2006.

Fisker, Anna Marie. *Food and Architecture* (Mad og Arkitektur). Phd. dissertation, Aalborg University, Department of Architecture & Design. Aalborg, 2003.

Det Danske Kulturministerium [The Danish Ministry of Culture]. *Kulturkanon* [Cultural Canons]. Det Danske Kulturministerie, Copenhagen, 2007.

Strong, Roy. *Feast, a history of grand eating*. London: Pimlico, 2002.

Research interviews

Jan Ringmose, antiquarian: http://jamerantik.dk/flora_danica.htm

Lise Nönnecke, Former painter and employed at Royal Copenhagen for many years. She now works in a shop in Amager, Copenhagen.

An Edible Wild Thistle from the Lebanese Mountains

Anissa Helou

Gundelia Tournefortii – *Tournefort's gundelia:* akkub *or* kardi *in conversational Arabic, as well as* kankar; akuvit ha-galgal *in Hebrew..*

Wild edible plants are, or to be more accurate were, an important part of the Lebanese diet. *Bi-sallqoh* is the Arab expression describing people going out into the wild (*barriyeh*) to pick all kinds of wild greens and plants such as mallow (*khebbayzeh*), yellow wood sorrel (*hommaydah*), borage (*l'san el-thor*), hawthorn (*za'rur*), and tumbleweed (*'akkub*). This activity takes place in spring and summer and the term comes from the word *sliq* used to describe the whole group of wild plants.

Tumbleweed is the common name for *akkub*, while the Latin name is *Gundelia Tournefortii,* Tournefort's gundelia. The plant is a member of the Aster family (Asteracea or Compositae) and it is related to thistles and artichokes. It became famous in 1998 when its pollen grains were found on the Shroud of Turin but it has been in use since ancient times – it is mentioned in the Babylonian Talmud and the Bible.

Akkub is a highly prized seasonal delicacy not only in Lebanon, but also in Syria, Jordan, and Israel. It grows in the plains and in the mountains, up to 3000 metres. It requires a sunny position and a well-drained soil. At the lower altitudes, it grows on the waste grounds of limestone hills of Jordan and in the semi-desert areas near Jerusalem.

The Hebrew name, *akuvit ha-galgal*, combines the Talmudic name for the plant, *akkub,* which is still the plant's modern Arabic name, with the biblical *galgal.*

An Edible Wild Thistle from the Lebanese Mountains

The preparation of *akkub* is very time-consuming, and rather perilous. The prickly parts need to be peeled to expose the bloom and stalks, which are then cooked, pretty much like artichoke, although the taste of *akkub* is a cross between asparagus and artichoke. *Akkub* is used in stews and soups; it is also fried and made into *kibbeh*.

When *akkub* is prepared in a stew (*yakhnet akkub*), it is cooked with minced meat and onion. The meat and onions are fried in olive oil before the *akkub* is added together with a little flour, to thicken the sauce. Once the *akkub* is tender, lemon juice is added and the stew is served with rice. For *kibbeh*, the stalks are boiled and ground, then mixed with fine burghul and grated onion and seasoned with salt, pepper, and the Lebanese seven-spice mixture. The *kibbeh* is then shaped into balls and fried, or it can be made *bil-saniyeh,* that is, made into a pie, with a walnut and onion filling seasoned with pomegranate syrup.

Some people freeze *akkub* after they clean it to use out of season, while others preserve it in brine. Freezing *akkub* produces better results than preserving it in brine, which makes it too soft and rather salty. *Akkub* is not only delicious but also nutritious and healthy. The plant is high in protein, iron, and vitamins A and B. It is said to improve the appetite as well as strengthen the blood and nervous system. It also has anti-cancer properties and is good for kidney problems.

Akkub is in danger of becoming depleted, given that it grows wild but is sold commercially. It propagates itself by rolling in the wind and dispersing its seeds. At the end of the season, the plant's stem separates from the root. This allows the entire plant, which is round and rolls like a ball, to be carried by the wind, and as it is driven by the wind, it disperses its seeds on steppe and field. (*Galgal* also means wheel in Hebrew; the plant's name probably derived from its habit of rolling across the fields like a wheel.) The dispersal of *akkub* takes place at about the same time as the wheat harvest as indicated by the prophet Isaiah (17:13) – 'driven before the wind like chaff on the hills, like tumbleweed (*galgal*) before a gale.'

Recently a programme was initiated by Dr Batal, when he was still at the American University of Beirut, to encourage farmers to grow *akkub* either near their houses or on the edge of fields where they plant other crops. According to Batal: 'Some plants, like *akkub*, for instance, which is harvested and exported, might be in danger in the future if they are too heavily marketed.' The project also allows for training women to use and market wild edible plants like *akkub*, and to train young people to recognize wild edible plants and their health benefits as well as to cook and consume them.

The American University of Beirut has a Food Heritage Foundation, headed by Professor Shadi Hamadeh, which is doing a lot of work on preserving Lebanon's food heritage, of which wild edible plants are an important part.

Luckily, *akkub* is not yet a rare nor a luxurious foodstuff and if you happen to be in Lebanon in season, you will not have a problem finding it in farmers' markets or at greengrocers in the market. I have also seen it sold in Syria, by families who bring their produce to a street corner and stay there until they sell whatever they have.

Allotment Diaries

Phil Iddison

Introduction

I am now in my fifth year as an allotment-holder in West London.[1] The primary purpose of an allotment is to provide family food, be it vegetables, fruit or small livestock. My personal aims are to provide the majority of our vegetables, and to a lesser extent, fruit, year round with an emphasis on variety and quality.

In October 2003 my first plot of five rods was covered in a mix of weeds and soft fruit canes.[2] By the start of 2007, the first plot was in full production and during the year I took over an adjacent plot of similar size.

I have kept various forms of diary to record events and the produce from my allotment. The principal document is the annual record of planting and produce that I have kept on a spreadsheet since the start of harvesting in 2004. I have also intermittently kept a narrative diary that has become more formalized since the winter of 2007/8. I have also prepared layout plans to record where the crops have been located each year and other incidental information. Finally I have taken many digital photos that form a visual diary of the allotment through the seasons.

In 2007 the combined area of about 250 square metres provided over 360 kilograms of vegetables out of a total production of 460 kilograms of food. I define the vegetable kingdom based on the most common end-use rather than a strict botanical definition. Vegetables are savoury and fruit sweet; for instance, I include tomatoes, peppers, beans, and squashes as vegetables but exclude rhubarb.

Based on supermarket prices I valued the vegetable crop at £650 for a total outlay of £250 in plot rental, seeds, compost, tools, and sundries. This does not take into account the value of the fruit crops, which are relatively far more valuable than the vegetables.

I grew 33 different vegetables and a total of 54 different cultivars. The table overleaf details the breakdown of the 2007 vegetable harvest.

My vegetable production is not organic. I use chemicals sparingly and, more importantly, I know precisely which chemicals have been used, on which plants, and when they were used. True organic production may not be achievable; some cultivation problems are intractable without intervention using selected chemicals. Most of the waste from the allotment is composted or burnt for potash.

Some vegetables are given away to family and friends but the majority go immediately into our own kitchen or to various forms of storage. Onions and garlic are tied into plaits, potatoes go into old pillowcases inside a cardboard box in the shed, and squashes occupy the spare bedroom at home. Only a small part of the bean crop is frozen.

We are not self-sufficient, but I estimate that we produce at least two-thirds of our

VEGETABLES HARVESTED IN 2007			
CROP	HARVEST (Kg)	CROP	HARVEST (Kg)
AUBERGINE	3.13	ONION	8.45
BEETROOT	11.61	PAK CHOI	3.99
BORLOTTI BEANS	7.80	PARSLEY	0.03
BROCCOLI	6.22	PARSNIP	6.86
BUNCHING ONION	0.16	PEAS	2.43
BROAD BEANS	15.53	PEPPERS	3.68
CARROT	22.07	POTATO	129.07
CAULIFLOWER	9.04	RADISH	1.78
CELERY	0.79	ROCAMBOLE	0.07
CHARD	6.38	RUNNER BEANS	2.83
CHILLIS	2.31	SAGE	0.89
COURGETTE	15.76	SALSIFY	0.87
CUCUMBER	3.95	SAVOY CABBAGE	2.77
FENNEL	1.20	SHALLOT	2.63
FRENCH BEANS	11.6	SQUASH/PUMPKIN	50.79
HORSERADISH	1.68	SWEETCORN	10.49
GARLIC	0.83	TOMATO	1.76
KALE	3.49	TURNIPS	0.30
LEEK	6.77	WINTER GREENS	2.29
LETTUCE	1.44		
		TOTAL	363.74 Kg

own vegetable consumption. We ate our own potatoes for 324 days of the last year, onions for 256 days, garlic for 335 days, carrots for 278 days, and broccoli for 291 days. In contrast some crops have short seasons as a fresh vegetable; I was picking broad beans for only 34 days.

 Productivity compared to commercial production is difficult to assess because the allotment crops are so varied and often in quite small quantities. However, last year all the potatoes were grown in a concentrated area measuring 5 metres by 5.5 metres. Commercial potato yields vary between 3.6 and 6.3 kilograms per square metre. The upper bound figure is only achieved with substantial chemical fertilizer application. My yield was a very respectable 5.4 kilograms per square metre and was largely organic.

2007 vegetable production

Production fluctuates during the year. Late winter and early spring are a lean time for fresh vegetables whilst July to September see the harvesting of the main storage crops such as potatoes and squashes. The monthly harvest shown in the following table for the first five years demonstrates these points.

MONTHLY VEGETABLE HARVEST (Kg)					
YEAR	2004	2005	2006	2007	2008
MONTH					
JANUARY	0	7.428	0.47	3.706	14.151
FEBRUARY	0	3.995	0.708	2.384	13.67
MARCH	0	0.2	0.01	2.93	12.924
APRIL	0	0.276	0.02	3.81	7.594
MAY	0	0.854	0.78	4.802	6.475
JUNE	1.854	9.45	7.554	42.953	32.558
JULY	13.024	27.671	39.365	51.799	48.017
AUGUST	30.78	79.373	105.385	136.813	148.096
SEPTEMBER	34.614	54.284	121.317	62.989	125.048
OCTOBER	34.667	33.392	14.814	23.815	21.579
NOVEMBER	11.316	6.154	9.486	16.014	0
DECEMBER	1.886	1.866	7.632	12.204	0
TOTAL	128.141	224.943	307.541	364.219	408.533

The data also show that my aim to spread production throughout the year is having some success. During the first winters we still relied on commercial vegetables. Last winter it was satisfying to be able to bring home fresh green vegetables to augment the stored root crops, squashes and dried beans. This improvement is shown on the graph overleaf.

The planting plan for 2008 started out with a total of 83 different varieties of vegetables from 46 species in nine major families. With this selection I am possibly matching my local supermarket on total choice, if not on seasonal choice. I can select vegetables for propagation that do not have a commercial market for such reasons as fragility or lack of demand. The chart overleaf indicates the breakdown of the vegetable crops by family.

Allotment characteristics

The principal aspects of an allotment supply of vegetables are choice, control, seasonality, and freshness. The allotment becomes your own greengrocer. The allotment

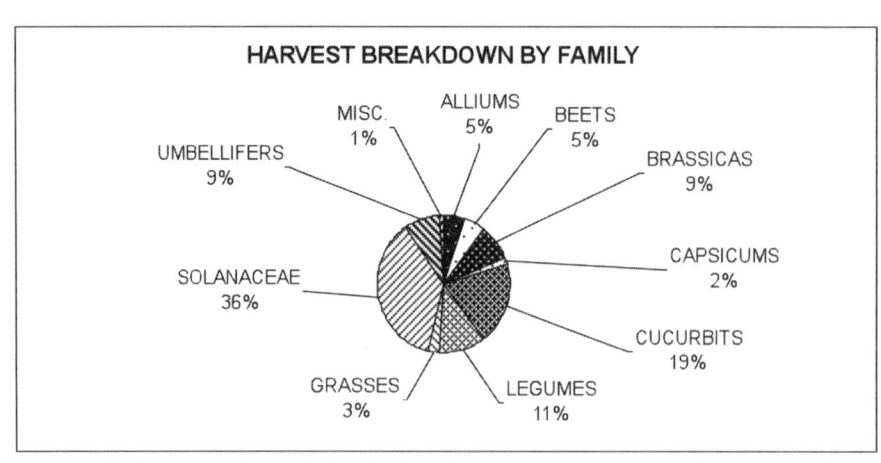

also provides space to experiment with different crops and cultivars. Chance provides culinary challenges and opportunities.

This is not achieved overnight, in a season, or even in a number of years. The process is continuous and, one hopes, incremental, but it does have setbacks. Perfection is rarely achieved but satisfaction comes in many forms, from the visual delight of a developing radicchio head through the aroma of crushed fennel foliage to the hot flavour of red leaf mustard.

Processes can become cyclical. For instance: from the purchase of red leaf mustard seed, through sowing, nurturing seedlings, planting out, watering and protecting, harvest and consumption. In my first year with this crop I let a few excess plants flower in the following spring. They were quite a spectacle and produced seed to continue the process. This was particularly satisfying in the context of achieving sustainability.[2]

The plots are open plan. The site is like an open handbook through which you can browse and even interact in real time if the plot holders are around. You see your neighbour's efforts and crops, successes and failures, unlike the closed world of the private house garden. There is competition, support, advice, and sharing from fellow plot holders. I have introduced my traditionally-minded neighbours to new vegetables and unusual cultivars. In return I have received the benefits of sound advice from their long experience. There is a true community spirit based on common interests and aims.

Perfection

We are now used to purchasing perfect vegetables. An allotment does not deliver perfect vegetables, despite the media image projected in a gardening programme or the illustration in the seed catalogue of the end product.

Many of the vegetables that I collect would not be saleable or considered acceptable by the modern consumer, a terrible reflection on our modern expectation of perfection that is subtly emphasized in the media. It brings home the massive waste associated with commercial production and marketing. Mini vegetables are a trendy restaurant offering but they are also the by-product of my allotment crops. It is worth boiling the beetroot that are only a centimetre diameter. I find these tucked in amongst the more commercially sized roots. They are the product of my lack of patience to thin out the seedlings as they develop.

Aims

My aims are selective and different to those of a commercial cultivator:

o   Flavour is more important than appearance. An old variety with character and good flavour is more valued than the modern F1 hybrid selected to deliver good looks and bulk.
o   Minimizing waste. Therefore nothing gets thrown away because it is too small or

blemished. Picking the broad beans starts when the pods are at the mange tout stage, continues through beans the size of a full grown pea, and continues to the point at season end when the beans have to be shucked out of their tough outer skins to be enjoyed. Putting the effort into growing the crops makes me loathe to waste anything that is edible.
- o Accepting that pests exist. Therefore you share your Savoy cabbage or radicchio with the slugs or white fly. Some leaves will be perforated and the slugs have to be removed but this preferable to speculating how a farmer has managed to produce a perfect pest-free specimen.
- o Freshness. The broad bean pods can be ready to eat in an hour or two; the first peas are eaten straight from the pod on site.

Observation

Tending my allotment has led to the following observations to contrast my efforts to those of the commercial growers:

- o Extended seasons. For instance, the carrot crop starts in June with thinnings that only need a wash and ends in mid-winter with monsters that need editing and a good long stew. Commercial growers lift the entire crop at the point when it will give maximum economic return. The beetroot crop in 2007 yielded an extended harvesting period of 240 days from June to February 2008, a staple food that could be kept at the allotment like a larder.
- o Seasonality has become a hot topic. It is possible to extend natural seasons at the allotment by protecting plants with various simple measures such as cloches. Wider availability of cold weather crops such as mustards, kales etc has greatly improved the diversity of crops that can be grown through the winter to fill the hunger gap. The allotment also acts as a larder, keeping leaf and root crops in good condition to be harvested as needed. Having said that, there is a delight every year when a particular favourite comes into production. Peas and beans are the obvious examples but any vegetable that can be eaten raw is usually sampled as soon as it is harvested. I hope that I never become blasé about the taste of a fresh, crisp baby carrot.
- o Challenges. My first attempt at growing celeriac was an abject failure resulting in golf ball-sized roots. This year I have read up on the subject and am trialling two different cultivars. It will be interesting to see if the claims on the seed packet are realized.
- o Experimentation. I can afford to try different vegetables in small quantities; I do not have to sell my production. Commercial growers may be experimenting but there is minimal access to the results.
- o Opportunities. Access to a vegetable supply of your own choice combined with the vagaries of plant growth and productivity lead to opportunities to experiment with food. For instance, last winter our Sutherland kale, an old crofter's variety, provided 'cut

and come again' leaves as a young plant, followed by main crop winter greens, and in the spring produced edible flower shoots as the plants went into reproduction mode.
o Awareness of the realities of vegetable production. Crops can fail or produce minor quantities of vegetables. Each female squash flower has the chance to produce a magnificent fruit or may not be pollinated and just shrivel and die.

My appreciation of the seasons is sharpened through involvement in my own vegetable production. I used to select vegetables in the supermarket to meet desired menus, regardless of season. I am now quite content to wait until production starts at the allotment, enjoy the best of the fresh vegetables, and then wait for next year when we have eaten our own crop.

Diary extracts

Some of the points made above are illustrated with extracts from my diaries and harvest records.

| 7 July 2007 | Green tomatoes | 44 gm | all that was rescued from blighted plants |

Last year was disastrous for the tomato crop. The plants at the allotment were destroyed by tomato blight, a fungal disease that was rampant in the wet early summer weather. The plants had to be destroyed and burnt to minimize the chances of a repetition this year. The few green tomatoes that were rescued went into chutney.

| 3 August 2007 | Buttercup squash thinnings | 254 gm | Fried with a balsamic vinegar dressing |

The Buttercup squash plants were rampant and had to be cut back to stop them overwhelming adjacent plants. Immature fruit were saved from the thinnings and were fried in olive oil, given a balsamic vinegar dressing, and eaten as a hot mezze.

| 23 February 2008 | Chard plants | 400 gm | Whole plants cleared including the root |

Chard is a versatile cultivar of the beetroot family that is rarely available commercially in the UK. It develops a substantial tap root. Is this root edible? I boiled the roots and tried them. The answer is no, for although the root does have the characteristic earthy taste of beetroot, it does not have the sweetness and is packed with fibre to support the substantial top growth. It could only be famine food.

| 20 June 2008 | Chard leaves | 610 gm | Used in a yufka pie |

The first verbal recipe that I collected whilst in Istanbul was prompted by finding a *yufkaci*, a maker of the traditional thin unleavened bread used for *börek*. A cobbled back street in Anadolu Hisar led to the simple factory that produced the thin round sheets of part-cooked dough, 50 centimetres in diameter. The recipe called for a filling of cooked spinach, white cheese, eggs, cumin, and pastirma.[3] This was layered in the *yufka* with a basting and glaze of yoghurt, oil, and beaten egg and then baked in the oven. The chard plants had been left in the ground over the winter and as biennials they were growing six foot flower spikes. There were enough small leaf bracts to replace the spinach for the pie.

| 17 August 2008 | King Edward potatoes | 20.47 kg | Lot of small spuds, look for new recipes |

My first trial of the King Edward potato variety produced a good crop but there were a lot of small specimens about the size of a 50 pence coin. I did not want to waste them and searched for an appropriate recipe. Rather neatly, Sam & Sam Clark's recent book *Moro East,* based on the experience of their allotment in East London provided an ideal recipe, Hassan's cracked potatoes with coriander. The small raw potatoes cracked easily. Fried gently in olive oil with a generous amount of crushed coriander seed and finished with a glass of red wine they were delicious.

| 18 October 2008 | Borlotti beans + others | 762 gm | For drying |

I always plant a selection of climbing beans in a long row. This year I planted borlotti beans for drying: a mix of green, purple, and yellow French beans to eat fresh and also some experimental golden bean seed from a neighbour that had originally come from Madeira. I had saved the green and purple French bean seed from the previous year. The results were rather mixed. There appeared to be some hybridization between the saved seed, which resulted in beans of varying hues; however, all were perfectly edible as fresh beans. As the season progressed the bean pods were drying on the vines and each harvest produced a mix of dried beans. A selection of the freshly dried seed was used in a bean stew with chorizo.

Future developments

Demand for allotments is high at a time when they are under serious pressure from development, particularly in cities.[4] Awakening interest in the quality of food and its source is combined with regular media references to allotments.

Allotment Diaries

An allotment requires intense and sustained effort to bring it into full production. The general rule is that it takes two hours of work per rod per week. A five-rod plot therefore requires 500 hours input per annum, about a third of the time spent at a full-time job. Few people have this amount of time to spare in busy urban lifestyles with many competing interests. I kept a diary record of my time spent on the allotment this year and am averaging 1.3 hours per rod per week. I guess that I work efficiently!

New allotment holders arrive full of enthusiasm and energy to create their vision of vegetable heaven. I have watched several new plot owners put in a few weekends' hard toil but then never return. They have paid rent for a year and in that year the plot heads to dereliction. This cycle may be repeated as another soul on the waiting list tests their stamina or takes short cuts and is overcome by the natural fecundity of the weed population.

It all looks so easy in the media bites of the TV gardening programme. A five-minute slot in a weekly programme seems to be all that is required to feed the whole family. Few people these days have had the role model of a grandfather or father who devoted their leisure hours to tending a vegetable plot whilst inculcating their knowledge in the future generations. I am intrigued when friends visit the allotment and do not recognize vegetable plants in their natural state. They are at a loss when faced with a vista of potato plants in flower.

There are plenty of books and articles providing the technical knowledge on vegetables and all the processes and techniques involved in successfully creating a vegetable allotment. Good practical advice is available from neighbours together with demonstrations if needed.

What is lacking is the commitment to hard effort and patience to await the rewards. Radishes will crop in 28 days from planting seed but over-wintering broad beans are in the ground for six months before a bean can be picked. A magazine headline or book title might exhort no-dig cultivation but the reality is that five rods of derelict land will take at least the combined weekends of a whole winter to dig over.

Some of the new plot holders do stick it out for that important first year and reap the rewards of the combination of their effort, time and nature.

On our allotment site there have been experiments with sub-division of plots, even down to the creation of hobby plots of only six square metres. Whilst this size might be more manageable it will not produce significant quantities of food.

What I do find heartening is the young families who have a plot. Toddlers and sub-teens are exposed to the routines of cultivation and enjoy an active outdoor life. I was there fifty years ago learning from both my grandfathers. I hope that some of these children get the opportunity in the future to have the enjoyment of an allotment.

Allotment Diaries

Notes

1. Allotments have a long history in Britain which can be traced in some form back to the middle ages. The tradition gained particular momentum and organization as a result of the Enclosure Acts and the industrial revolution which precipitated major changes in both rural and city life. My allotment is on a site next to Bushy Park that was established in 1896. There are over 400 plots on the site which I estimate produces about 60 tons of food each year for the allotment holders.
2. My father did not buy any broad bean seeds for nearly forty years starting in the 1950s. He saved and replanted his own seed. There were two distinct bean types, a smaller green seed and a large white one. At the end of each season the last bean pods were left to dry on the bean haulms. The seeds were kept in an old Tate & Lyle Golden Syrup tin for use next year.
3. Air dried beef fillet coated in a paste of garlic, cumin, and chilli pepper.
4. During the Second World War allotments played an important role in feeding the nation. At the end of the war there were 1.4 million allotments. Now there are less than 300,000. See the paper by Lesley Acton, above.

Salvation in Sweetness?
Sugar Beets in Antebellum America

Cathy K. Kaufman

The sugar beet, *Beta vulgaris* var. *esculenta*, is a rather plain-looking beige root, averaging just over one and one-half feet long. Used as a food and fodder crop since at least the ancient world, it is distinguished by a high sucrose content ranging from 6 per cent in older varieties to nearly 20 per cent in modern cultivars. The American cookery writer Amelia Simmons, comparing the red and white beet, warned that 'the *white* has a sickish sweetness, which is disliked by many.'[1] The sugar beet was a minor crop until the turn of the nineteenth century, when the potential to extract table sugar from the root turned this disfavored vegetable into the hottest agricultural story of the time, first in Europe, and then in the United States. But more than just a scientific marvel, the sugar beet was enveloped in an utopian aura. Some nineteenth-century American abolitionists saw it as the ultimate weapon in the battle against slavery: cheap beet sugar would make slave-produced cane sugar uneconomical. These dreams for the peaceful withering away of the plantation system would come to naught, as the sugar beet industry only become commercially viable in 1870s America, after her bloody Civil War had done away with slavery and shattered the economic and social foundations of the Confederacy. The story of the sugar beet is a cautionary tale for current debates in America on the uncomfortable balance among capitalistic markets, governmental subsidies and tax policy for favored and powerful industries, and governmental support of research and development that could lead to new technologies that would change the economics of those entrenched industries.

The die was cast for the importance of the sugar beet nearly 150 years before scientists first started experimenting with them in the laboratory. The story is well known of how the rhythms of daily life in Europe shifted with the introduction of the fashionable drinks of coffee, tea, and chocolate. Enjoyed as expensive luxuries by the well-to-do in the late sixteenth and seventeenth centuries, these refreshments became a virtual necessity to all ranks over the next 200 years. By the nineteenth century, the working classes in France, England, and the United States routinely consumed coffee and tea as part of their cultural identity. Most of these 'hot liquor' aficionados heavily sweetened the bitter brews with sugar manufactured from one of several varieties of tropical sugar cane, grown either in Asia or in colonial outposts in the western hemisphere. This skyrocketing consumption of sugar in the age of Enlightenment and early industrialization fed one of the ugliest blots on history: the Triangle Trade's trafficking in slaves to work the cane plantations in the West Indies and American South.

In addition to seeking ever-cheaper sources for sweeteners, the battling among various colonial empires meant that the imported supplies of sugar were subject to the vagaries of trade, politics, war, and nature. Embargoes, blockades, and shipwrecks all could curtail the supply of sugar reaching a given destination, and tariffs and duties inflated the price for those not controlling the importation of this highly desired commodity. Starting in the mid-eighteenth century, European scientists began to experiment with extracting sugars from plants that could be cultivated locally. Not surprisingly, the first experiments were in non-maritime powers. In 1747, the pharmacist Andreas Sigismund Marggraf presented a paper to Academy of Berlin entitled 'Chemical Experiments, made with a view to extract genuine Sugar from several Plants which grow in these Countries.' Marggraf had managed to create sucrose from beets, carrots, and other root vegetables by heating the dried roots in alcohol; upon cooling, an unrefined form of sugar slowly precipitated out of the solution. His finished product was too crude, and the technology too inefficient, to manufacture commercially acceptable sugar, but the basic science was sound. His student, Franz Karl Achard, continued the experiments when funding from the Prussian government could be had. By the end of the eighteenth century, Achard published findings in the *Annales de Chimie* suggested that sugar could be manufactured from beets at a cost lower than purchasing imported cane.[2]

Some French scientists were eager to pursue Achard's research. Having lost many of its sugar-producing colonies in the West Indies during the Seven Years' War and thereafter having abolished slavery (albeit temporarily) in the remaining colonies as part of the *mentalité* of the French Revolution, demand was high for cheaper sugar than what the French could purchase in the world market, especially given Napoleon's war with Britain. Private investors, however, hearing rumors that peace was at hand, feared that the price of cane sugar would drop and were hesitant to risk significant capitol in the unproven beet technology. Thus, at the turn of the nineteenth century, only a token effort to perfect beet sugar refinery was made in two small factories near Paris. The results were disappointing, possibly explained by poor quality beets and unskilled labor.[3]

Interest in beet sugar would revive in 1806, when Napoleon instituted the Continental System, which attempted to bar British trade with France's allies and the areas under her control. Colonial sugar was one of the goods in extremely short supply, although smugglers occasionally penetrated the permeable European coastline and some nations refused to join the boycott. Nonetheless, most French were sweetening their drinks with a form of treacle made from grape juice, considered a poor substitute for cane sugar. The very practical chemist Jean-Antoine Chaptal,[4] who was Napoleon's Minister of the Interior, revisited the Prussian research, directing further experiments to master refining the beet sugar through the innovative use of charcoal. After initial laboratory successes Napoleon was advised that beet sugar might be refined on an industrial scale. In 1811 and 1812, Napoleon issued a series of decrees designed to cease all trade in British colonial sugar, to dedicate significant French lands to the cultivation of the sugar beet, and to create six schools to teach 100 advanced science students the process of beet sugar refining. These graduates would work in one of four imperial manufactories, also to be funded by the State, with an anticipated annual processing capacity of 4.4 million pounds of raw sugar. Chaptal, as Minister of the Interior, also was given significant funds (approximately $200,000 in early nineteenth-century dollars) to encourage the nascent industry, and the government exempted beet sugar from duties for four years.[5]

The French got to work quickly and the results of the 1813 season appeared successful, although exactly how much sugar was produced and at what cost is never disclosed in the Anglophone press.[6] The industry lost governmental funding and temporarily collapsed after Napoleon's defeat at Waterloo. A few private manufactories continued, aided in one case by unusual luck and more generally by the substantial tariffs levied on imported cane sugar. The initial Napoleonic successes were remembered, however, and in the 1820s renewed efforts improved refining techniques. From the mid-1820s, France's sweet tooth was satisfied by an ever-increasing proportion of beet to cane sugar, with beet surpassing cane by about 1850. Belgium, Russia, Poland, Holland, Zollverein (essentially Germany), and Sweden all had beet sugar factories by 1865; 50 per cent of the sugar consumed in Europe by that date was from continental beet production.[7]

By contrast, the vast bulk of the sugar consumed in antebellum America was made from cane and was the product of slave labor, whether imported from the British West Indies before the final abolition of slavery there in 1840, or produced on slave-supported plantations in Louisiana. Slavery was a uniquely difficult conundrum for early Americans. The fact that slavery took place on the American homeland, rather than on distant colonies, raised immediate questions of what would happen if slaves were freed. Would they continue as wage laborers on the plantations, or migrate to the more industrialized North and compete for jobs? Was slavery economically necessary to the agrarian South? Was the beneficent paternalism of the 'kindly' master the most compassionate accommodation, at least until an appropriate transition economy was developed? Or was slavery so morally flawed that immediate abolition was the only option? Should freed Africans be returned to the newly founded Liberia to ease racial and economic tensions? Respected opinions could be found all along this spectrum.[8] Neither did religion give clear guidance: in the Scripture-quoting atmosphere of nineteenth-century America, each side could evoke Biblical passages to sanction its position, making questions of God's will fair grounds for debate. An anonymous passage from 1798 captures some of the competing concerns:

> I abhor the slave trade, [but] I neither conceive slaveholding inconsistent with my Conscience, nor derogatory to the Christian religion, the rights of mankind, or the different orders of subordination; yet [I] have no objection to its gradual abolition.[9]

Faint rumblings of opposition to slavery had started in colonial America while the Constitution was being drafted, but these were hardly sufficient to dismantle the pre-existing institution. The Founding Fathers, many of whom owned slaves, knew that the question was divisive and had no acceptable solutions. Many had a sense that the tide was slowly turning against slavery, but struck a grand bargain to obtain enough votes from the Southern states to ratify the Constitution: the importation of slaves would be protected only for 20 years, slaves would be banned in north-western territories, and slave-holding states would get congressional representation through the notorious three-fifths clause, which gave states additional Congressmen calculated on the number of slaves in each state. This bit of political legerdemain, while promising to staunch the flow of new slaves into the United States in the future, also made national abolition less likely through the political process alone.

While the political debates were playing out, abolitionist groups were formed in northern states to influence popular opinion against slavery, much like the public relations assaults of the British anti-saccharrites against slavery and slave products. Some of the first and most important American opponents of slavery were the Quakers. By the end of the eighteenth century, the Friends had manumitted all of their slaves and began looking for other ways to spur abolitionist sentiment. Philadelphia, home to a large proportion of Quakers, became a locus of abolitionist activity in the 1820s and thereafter. Believing that undermining the economies of the plantations by boycotting products produced by slaves would bring a voluntary end to slavery, the pacifist Quakers reminded consumers of the 'blood-stained' circumstances of slave-produced sugar production and urged consumers to decline slave sugar at their social teas.[10] The Ohio abolitionist paper, *The Genius of Universal Emancipation*, regularly published didactic poems in support of sugar boycotts, ranging from an attitude of smug, moral superiority:

> No, dear lady, none for me?
> Though squeamish some may think it,
> West India sugar spoils my tea;
> I cannot, dare not, drink it.[11]

to one suggesting that the consumption of cane sugar was tantamount to cannibalism:

> Oh press me not to taste again
> Of those luxurious banquet sweets!
> Or hide from view the dark red stain
> That still my shuddering vision meets.
>
> Away! 'Tis loathsome! Bear me hence!
> I cannot feed on human sighs,
> Or feast with sweets my palate's sense,
> While blood is 'neath the fair disguise.
> No, never let me taste again

> Of aught besides the coarsest fare,
> Far rather, than my conscience stain,
> With the polluted luxuries there.[12]

But, much like Napoleon, the Quakers recognized that difficulty mere mortals would have in foregoing sugar entirely. To sate the appetite for sugar and other fashionable goods wrought on the backs of slaves, the Quakers started Free Produce Societies, opening stores where consumers could exercise their political beliefs through their pocketbooks and purchase 'Free Labor' sugar brought in from East Asia, as well as other unsullied products. The problems with free sugar, as with most of the free labor products, were both quantitative and qualitative: free labor sugar was hard to come by, was more expensive than slave sugar, and worst of all, tasted awful. 'Free sugar,' rued one abolitionist, 'was not always as free from other taints as from slavery … and free candies [were] an abomination.'[13]

In seeking a solution to the problem of inferior Asian sugar, Americans of different political and religious stripes took notice of the increasing success of the French and German beet sugar industries. In 1836, the Beet-Sugar Society of Philadelphia was formed and sent James Pedder to France to study the process of beet cultivation and refining. While Pedder provided a detailed, optimistic report on the French industry, he did not experiment with the process, leaving it a theoretical ideal.[14] Also that year, David Lee Child independently visited France, Belgium, and Germany to glean the best practices with the plan of introducing a practical technology to farmers in the temperate United States.[15] Child was uniquely situated to popularize beet sugar, as he was a notorious celebrity in the 1830s, having served stints as a State legislator, newspaper editor, defendant in a highly-publicized libel trial, and all-around abolitionist agitator. He was also the husband of the best-selling young author Lydia Maria Child, whose work ran the gamut from novels to household manuals to a controversial exegesis on the evils of slavery that cost her a prestigious position as the founder and editor of the first successful American magazine for children, *The Juvenile Miscellany*.

Child had a reputation for fiscal irresponsibility, and the Childs were forced into a frugal existence notwithstanding their comfortable middle-class roots. However impractical Child may have been, he understood that he needed to make an economic, rather than moral, argument in favor of beet sugar. His 1840 monograph, *The Culture of the Beet and the Manufacture of Sugar*, summarized in painstaking detail his own experiments in refining beet sugar in Northampton, Massachusetts after returning from Europe so that a gentleman farmer could pursue this 'new and pleasant branch of rural economy.'[16] In reading *The Culture of the Beet*, one gets the sense of a man whose political and social agenda was carefully hidden in a thicket of what proved to be inadequate technology and bogus calculations. Much of *The Culture of the Beet* is spent explaining the refining process, complete with illustrations that should allow almost anyone with a minimal amount of capital to set up a beet sugar factory. The last third of the book gives an economic justification for switching to beet sugar from cane. Not only could the amount of sugar extracted be increased by the development of improved cultivars, but the beets performed a simultaneous benefit of returning nitrogen to the soil. By emphasizing the secondary uses for processed beets as fodder, which could increase cattle herds and the amount of manure available to put to other farming uses, Child argued that beet sugar refining could recoup its investment and turn a profit quickly. Only once towards the end of this 156-page treatise does he off-handedly mention beet sugar as an antidote to slavery.

The Culture of the Beet received wide notices in the specialized agricultural and abolitionist press. While the abolitionist papers wanted to promote the possibilities of beet sugar, reading between the lines, even Child's most ardent supporters sounded skeptical of the practicality of beet sugar. *The Emancipator*, the brainchild of activist William Lloyd Garrison, a close friend of the Childs, perhaps revealed too much when it touted Child's 'unqualified *belief* that sugar can be made in Massachusetts for 4 cents per lb. The actual success of these attempts must be a heavy blow at slavery in America; and for this, as well as other patriotic as well as personal reasons, *we hope he may succeed* to his utmost wishes' (emphasis added).[17] Other sectors of the agricultural press were less propagandistic and more pragmatic:

> Too much has been said of the manufacture of beet sugar in a small way to permit us to pass this topic in silence. There can be no doubt of the practicality of making beet sugar on any required scale from 1 lb. to 10,000 lbs. per day, but it cannot be made on any scale without considerable skill and experience…There is no efficient apparatus, which is simple and cheap enough to suit the purpose of common farmers. Yet we believe that the time will come when sugar-mills will be set up by the more forehanded of our farmers, as cider-mills are now… We doubt not [that Child's book] will be the means of promoting the sugar manufacture among us – a manufacture …[that] will also render us independent of foreign countries for an article of daily consumption and which is now reckoned as one of the necessaries of life.[18]

Child's Northampton experiments were the best known of the early sugar beet efforts, but many attempted to refine sugar as a cottage industry and to improve the quality of sugar beets over the next twenty-five years.[19] Among the most quixotic was a small notice in the *Genesee Farmer and Gardener's Journal* and reprinted in New York's *The Colored American*, a journal devoted to 'Colored Americans – to be looked on as their own, and devoted to their interests.'[20] The piece pointed to a simplified German experiment for extracting beet sugar that, ostensibly, could be used by anyone, for the process 'was as simple as that of making a cup of tea.' Beet roots were dried, pulverized, boiled in water, and then clarified with animal charcoal and lime. According to the notice,

This eminently successful experiment proves the fact, that the immensely important article of sugar can be manufactured by an intelligent farmer's household, as easily as bread, pies, or cakes, can be made and baked. A rich garden spot of a quarter of an acre can produce enough beets to make a thousand pounds of sugar; no heavy capital or incorporated company is required to carry on the business.[21]

Other early notices exuded a jingoist confidence that 'Yankee ingenuity' could simplify the process that the French had made overly complicated, promising a painless solution to the sugar question.[22]

The more serious sugar beet enthusiasts, mindful that the European industry had been subsidized in its research and development stages, assumed that adequate funding would be available to secure success. Federal funding was improbable because of the disproportionate Southern representation in Congress. Nonetheless, *The Genesee Farmer and Gardener's Journal* (a paper with abolitionist and nationalist sympathies), writing about the founding of the Philadelphia Beet-Sugar Society and its dispatch of Pedder to Europe, explicitly assumed that sufficient private capital would be available to support the trials. The article 'trust[s], that as there will be no want of capital, so the spirit with which the enterprise has been entered upon will continue, until success crowns their efforts,' and concludes with the rhetorical flourish that,

There can be no reason why … we should remain tributary to other nations for so essential an article of domestic consumption. Nature has placed in our hands the means of becoming in this respect really, as well as nominally, independent. It only remains for us to avail ourselves of the science and skill of the age so desirable a result. As citizens and farmers of the north, shall we hesitate?[23]

These assumptions were misplaced. When Child was unsuccessful, *The Boston Herald* (rightly) condemned his work as underfunded and lacking in 'practical' men. *The Herald* called for a mere $20,000 (a tiny fraction of Napoleon's support in 1811–12) to put the experiment on solid ground, as well as the active involvement of the Agricultural division of the Federal Patent Office, another hopeless quest in antebellum America.[24] Small-scale experiments throughout the North and Midwest continued, and failure upon failure ensued. Each succeeding experiment started from the naïve belief just a bit more money or care in the cultivation of beets would guarantee success. Most spectacular was the effort in the 1850s in Utah by the Church of Jesus Christ of the Latter-day Saints, and undertaken primarily to free the Mormons from the shackles of economic discrimination, rather than from an abolitionist motive. The Mormons were reasonably well capitalized (although not on the scale of the French subsidies) and, unlike other American efforts, imported heavy equipment from England that had been proven successful. The Mormons transported the cumbersome equipment overland great distances to Salt Lake City and bought top quality beet seed from France, only to be frustrated by the saline soil conditions that were hostile to the refining of flavorful sugar.

Not all abolitionists were lost in an idealistic miasma. The African-American newspaper, *The National Era*, reviewed the options for obtaining sugar without slave labor, or, in the anonymous author's graceful phrasing, 'to show that the sweet may be obtained without the bitter, and that there is no necessary connection between bondage and Muscovadoes.'[25] The article critiqued the quality of beet sugar, even if refined according to French standards and further dismissed the possibility of refining sugar as a cottage industry. It concluded that large-scale operations were the only efficient means to avoid fermentation of the sugar solution and resulting loss of quality and quantity. *The National Era's* solution to slavery was free trade with Asian sources, the elimination of duties, and the market would take care of the rest, a variation on the Quakers' Free Produce.

None of the pre-1870s efforts to reduce the theory of beet sugar refinery in America to practice was successful, notwithstanding the well-documented growth and profitability of the European industry. In 1869, the weekly *Scientific American* wrote that,

> Many causes are now at work to interest the capital of this country in the manufacture of beetroot sugar. Among these may be enumerated, first, the depression of the sugar trade of the West Indies consequent upon the competition of European beet sugar, which threatens to compel the abandonment of the business on many plantations. Second, the changed condition of affairs in the sugar growing districts of the United States on account of the abolishment of slavery and the increased cost of labor resulting therefrom… . [Based in the French example], it is we think impossible to show why this industry should not spring at once into healthy activity, if capitalists would open their eyes to the promise of profit it offers.[26]

Shortly after the *Scientific American* article was published, the first commercially successful factory for beet sugar refining was constructed in Alvarado, California, although the factory would not turn a profit until 1879. The swiftness with which American industry adapted and improved the European technology, the willingness to wait nearly a decade for profitability, and the vast public funding and subsidies that were deemed necessary to prop up the industry through the twentieth century, have parallels in contemporary debates about alternative energy sources and agricultural policy. Unfortunately for nineteenth-century Americans, neither sufficient public nor private resources were devoted to beet sugar experimentation and production in antebellum America to help undermine the economic basis of slavery. Whether a vigorous sugar beet industry could have avoided the inferno of the Civil War is unlikely, given the Confederacy's seceding from the Union. Nonetheless, an economically weakened South likely would have shortened the conflagration and reduced the utter devastation of the war. In any event, the development of the beet sugar industry was inevitable and relatively easy once capitalists saw the potential, and was unnecessarily delayed in the United States by the availability of cheap cane sugar produced by the plantation system. There might have been some salvation in sweetness, had public and private partnerships focused on factors beyond the immediate bottom line.

Notes

1. Simmons, p.11.
2. Child [1839], pp. 432–3.
3. Ibid.
4. Chaptal also invented the chaptalisation process of adding sugar to wine to increase the alcohol levels.
5. Decrees of March 26, 1811 and January 15, 1812, translated and reproduced in Truman A. Palmer, Beet Sugar: A Brief History of Its Origin and Development, Senate Report No. 204, 57th Congress, 2d Session (3/3/1903); Leonard J. Arrington. *Beet Sugar in the West: A History of the Utah-Idaho Sugar Company, 1891–1966.* University of Washington Press, 1966, pp. 4–5.
6. 'Glucose.' *The British Food Journal*, vol. 36, no. 10, p. 91 (October 1934); Child [1839], 436; *Boston Herald*, 2/12/1858, p. 2.
7. Grant, pp 12–22; *The Genesee Farmer and Gardener's Journal*, 2/2/1839, p. 38.
8. *See, e.g.,* Horace Greeley, 'The Results of Emancipation,' *The Independent*, 1/29/1863, p. 1.
9. Quoted in Finnie, p. 322.
10. *Friends' Review*, 4/17/1852, p. 491.
11. 'Gertrude,' The Use of Free Produce,' *The Genius of Universal Emancipation*, Jan. 1831, p. 155.
12. Quoted in Faulkner, p. 377.
13. Letters of Lucretia Moot, quoted in Nuermberger, p. 99.
14. Harris, p. 16; 'Observations on the Cultivation of the Sugar Beet.' *Farmer's Register*, 2/29/1840, p. 90.
15. 'Beet Sugar.' *Portsmouth Journal of Literature and Politics*, 1/18/1840, p. 2.
16. Child [1840], p.4.
17. 'Sugar Beet and Beet Sugar.' *The Emancipator*, 2/13/1840, p. 167.
18. 'The Culture of the Beet.' *Maine Farmer and Journal of the Useful Arts*, 3/14/1840, pp. 73–4.
19. *See* Grant, Appendix, pp. 144–158.
20. Samuel E. Cornish, ed., *Colored American*, 3/4/1837.
21. 'The Manufacture of Beet Sugar: The German Improvement Over the French.' *Colored American*, 11/17/1838. See also 'Beet Sugar.' *The Genesee Farmer*, 2/2/1839, p. 38.
22. 'Sugar Beet.' *The Maine Farmer and Journal of the Useful Arts*, 3/27/1838, p. 52.
23. 'Sugar from Beets.' *The Genesee Farmer*, 8/6/1836, p. 253.
24. 'The Manufacture of Beet Sugar.' *The Boston Herald*, 12/16/1856, p. 2
25. 'Free Labor vs. Slave Labor.' *The National Era*, 4/24/1856.
26. 'American Beet Root Sugar.' *Scientific American*, vol. XX, no. 4, p. 57, 1/23/1869.

Bibliography

Child, David Lee. *The Culture of the Beet and the Manufacture of Sugar.* Boston: Weeks & Jordan, 1840.
———. 'Beet Sugar Manufacture.' *North American Review*, 48:415–47 (April 1839).
Child, Lydia Maria. *An Appeal in Favor of that Class of Americans Called Africans* [1833]. (Carolyn L. Karcher, ed. and intro.). Amherst: University of Massachusetts Press, 1996.
Faulkner, Carol. 'The Root of the Evil: Free Produce and Radical Antislavery, 1820–1860.' *Journal of the*

Early Republic 27:377–405 (Fall 2007).

Finnie, Gordon E. 'The Antislavery Movement in the Upper South Before 1840.' *Journal of Southern History* 35:319–42 (Aug. 1969).

Garrison, Wendell Philips. 'Free Produce Among the Quakers.' *Atlantic Monthly* 22:485–94 (Oct. 1868).

Glickman, Lawrence B. '"Buy for the Sake of the Slave": Abolitionism and the Origins of American Consumer Activism.' *American Quarterly* 56:889–912 (Dec. 2004).

Grant, E.B. *Beet Root Sugar and the Cultivation of the Beet.* Boston: Lee & Shepard, 1867.

Harris, Franklin Stewart. *The Sugar Beet in America.* New York: Macmillan, 1919.

Jeffrey, Julie Roy. *The Great Silent Army of Abolitionism: Ordinary Women in the Antislavery Movement.* Chapel Hill: University of North Carolina Press, 1998.

Mintz, Sidney W. *Sweetness and Power: The Place of Sugar in Modern History.* New York: Penguin, 1985.

Nuermberger, Ruth Ketring. *The Free Produce Movement: A Quaker Protest Against Slavery.* Durham, North Carolina: Duke University Press, 1942.

Phillips, Ulrich B. 'Plantations with Free and Slave Labor.' *American Historical Review* 30:738–53 (Jul. 1925).

Simmons, Amelia. *American Cookery* [1758]. New York: Dover Facsimile Edition, 1958.

Wilkinson, E.C. '"Touch Not, Taste Not, Handle Not": The Abolitionist Debate Over the Free Produce Movement.' *Columbia Historical Review* 2:2–14 (Winter 2002).

Up on the Farm:
The Role of Vegetables in Conquering Space

Jane Levi

Having a good meal is, of course, something more than downing one's food. It's a complicated process combining physiological and psychological elements. Even in a short flight tasty, favourite dishes can provide the astronauts with relaxation during their strenuous work. Indeed, not only the taste of the food is important, but even the circumstances in which it is eaten. A spotless table-cloth, attractive plates, pleasant music, and enjoyable conversation help a person to rest while eating, whereas tasteless and unappetizing dishes, and even an unattractively laid table, can cause irritation and, far from being conducive to enjoying a meal, can retard the secretion of digestive juices.[1]

Yuri Gagarin had a clear opinion on the importance of food for astronauts and cosmonauts, and a glorious vision of how food and eating in space should be experienced. Sadly, the realities of space travel, especially during the 1960s and 1970s, proved to be a less savoury experience than he had foreseen.

Vegetable provision on the space menu helps us trace a change in attitude toward space food, in particular in the United States. Through vegetables we can trace a discernible shift in the accepted cultural norms, from the 1960s idea that space heroes really only eat meat; through the 1970s and 1980s where space colonists and people on space stations were expected to demand meat with some vegetables on the side; to the notion in the 1990s and early 2000s that long-term space travellers may need to be self-sufficient and largely vegetarian.

Regardless of the menu it is worth considering what it is really like in space, and how this affects health, appetite, and eating. Between 44 and 67 per cent of space travellers suffer from the loss of appetite, vomiting, nausea, and stomach awareness induced by space motion sickness or SMS.[2] Significant shifts of fluid to the head impact the senses of smell and taste.[3] Muscles and bones atrophy, the effect reduced but not stopped by the considerable effort of resistance training in the challenging conditions of microgravity. There is a risk of cardiac arrhythmia and near certainty of cardiovascular de-conditioning; red blood cells are lost; and exposure to high levels of radiation increases the propensity to develop cataracts in later life as well as increase the chances of developing other radiation-related diseases such as cancer.[4] Digestion isn't easy; flatulence an occupational hazard.[5] Conditions are cramped and stuffy but you can't ever open a window. In astonishingly stressful circumstances crews not only have to keep the vessel afloat and face the dangerous journey back through the Earth's fiery atmosphere to a safe landing; they have to do their real job whilst there, conducting complex experiments to tight deadlines.

Put yourself in that situation and consider your dinner. How would you feel about food? What would you want to eat, if you could face it at all? Something light; easily digestible; stimulating to the eye; fresh and colourful? Or some comfort food, stodgy and familiar?

In the early years of space exploration it was not consideration of appetite but the approach taken to provisioning that had the strongest impact on the menu. Both the USSR and the USA adopted technologies originally developed for high-altitude pilots – the origin of now infamous tubes of puréed foods, frequently fruits or vegetables. Beyond this, though, the approach between the two nations diverged, for cultural as well as practical reasons, reflected in their attitude to vegetables.

The USSR tended to work with existing technologies, using canned foods (such as potatoes and other vegetables) and dried foods (such as vegetable soups like borsch). This practical approach not only ensured that much of the food taken into space was immediately recognizable as food to the cosmonauts, and might even be the same as some of what they ate at home; it also meant that many more specific vegetables, like beans or mushrooms, were made available in forms other than purées, incidental ingredients in meat dishes, or soups.

The Soviets had started their programme with a clear intention to establish long term habitation in space – their first inhabited space station was launched in 1971 – and the planners took into account psychological factors, including food and eating, much earlier than their US counterparts. A table for eating as well as working was provided on board Mir in recognition of the importance of conviviality. This recognition of the psychological importance of food and in particular vegetables extended to the experiments conducted on board the space station. The beneficial effect of watching plants grow was noted: 'It turned out that watching a plant grow in space was therapeutic, even for cosmonauts with no previous interest in horticulture.'[6] The plants in question also provided a ready and eagerly seized upon source of fresh food. Onions and radishes were frequently grown experimentally, and on several occasions the crew found the crop irresistible and ate it. Sometimes a celebration provided the excuse: 'Sevastyanov's 40th birthday was on 8 July, and Klimuk's 33rd occurred two days later. They celebrated with a feast of spring onions which had grown in the Oasis [the plant propagation unit].'[7] Sometimes these fresh feasts were authorized, as during a long mission in 1987: 'As a special treat, the cosmonauts were told that they could eat the radishes and onions that they had grown.'[8] Occasionally, human greed took over and the arrival of fresh pants was too much for the crew: 'Progress 14 docked on 12 July. It delivered miniature onions for the cultivator, but they were so tempting that they were eaten instead.'[9] Whatever the circumstances, the urge to consume fresh, strongly flavoured vegetation after months in space eating the blander food from packages and cans was too strong to deny.

The Role of Vegetables in Conquering Space

For the US space programme, at least in the initial stages, food was wrapped into the great futuristic technological drive that the entire programme represented, often with unfortunate results. There was an obsession with the amount of space available to store anything, particularly food: what it would weigh; how long it would take to prepare; how much waste there would be. There was also coy reference to the undesirable digestive effects of pulses and too much fibre. The astronauts' priority was to work, and food and its transport through the body were an inconvenience: the administrators expected astronauts to treat journeys into space as camping trips, and to tolerate the many discomforts involved, including the poor food.[10] Cubes coated in various gums that would dissolve in saliva in the mouth to deliver a taste of toast and butter or fruit cake were developed, and eventually abandoned when they gave astronauts stomach upsets. No thought was given to how revolting it might be to have only cold water to rehydrate dried foods: one of Gemini's major discoveries was that astronauts don't like cold potatoes.[11] Early freeze drying techniques had other unpleasant effects. Oxygen left in the voids of the food can lead to oxidation, causing browning and deterioration, particularly in caretenoids. 'Carrots are especially susceptible to loss of color, loss of flavour and the development of "perfume-like" odor. Other vegetables develop a "hay-like" odor as a result of oxidation.'[12] None of these effects made the consumption of vegetables more appealing, and probably ensured that they remained in a minor role. They appeared in the diet as less important stew ingredients or as a small side dish for a meat main course, not as main dishes in their own right.

Although it was several years before any significant complaints were made, it became clear that the diet of Gemini and Apollo during the late 1960s was leaving astronauts dissatisfied. They lost weight, and they returned to Earth with far more uneaten food that they should have done. Still more worrying, they were not consuming the required nutrients. Food therefore became a major focus for the US programme for the first time when Skylab went into operation in the 1970s. Not only did the space station have a kind of dining area (designed by Raymond Loewy) leading to more structured meals and meal-times; the astronauts' food consumption was closely monitored and they were actually forced to eat their meals as required by the nutritionists. Although these crews were the first not to lose weight in space this wasn't always popular. Vegetables, which became a key menu item monitored from earth, came in for particular criticism. 'I found that if I reconstituted the peas, the beans, and the asparagus early, and then reheated them,' Pete Conrad quipped, 'I still didn't like them, but they were a lot easier to choke down than when I added the hot water, shook up the bag and then tried to get them down.'[13] The problems were compounded by an issue with the water pressurization system, which meant it was full of air bubbles, so that the astronauts could not help ingesting excessive air with their rehydrated food, leading inevitably to flatulence. As one astronaut commented, 'Farting about 500 times a day is not a way to go.'[14]

Matching the urges of the onion- and radish-eating cosmonauts, fresh vegetables featured as an important element in astronaut fantasies about homecoming. It could be argued that this reflects a natural craving for variety in both texture and flavour, as well as a desire for freshness. The sense of taste is affected by space travel, and almost all astronauts and cosmonauts develop a preference for strong condiments to enhance their food – for Russians garlic and for Americans chilli sauce. In a series of letters home from Mir Jerry Lineger puts 'a big, fresh salad' in the top four items for his first meal when he arrives home, adding, 'To be honest, anything would be fine, even John's baby food leftovers, as long as it doesn't come in a tube!'[15] Despite the technical team's feeling that space heroes would want and need protein in the form of meat, the experience of many of the crew members bears out the idea that more variety, more flavour and, in particular, more vegetables, whether whole or as flavour-enhancing garnish, were actually welcome.

Whilst reality unfolded on Skylab and Mir during the 1970s and early 1980s, another group of scientists was working in parallel on the practicalities of establishing space colonies. Some saw such extra-terrestrial escape as the obvious, even inevitable, next step for humankind, having exhausted and over-populated the earth. Several physicists' ideas (e.g. Dyson, Bernal) were developed into working designs for huge space stations that could support significant human communities. They would use the power of a star or the sun to remain in space and to support the energy needs of the population living, working, and farming inside. Solar reflection panels would move to simulate day and night for crop growth, and CO_2 condensers would ensure a plentiful supply of water. The descriptions and in particular the illustrations of these space communities are striking for their focus on green spaces for leisure as well as farming. Many interior views show green fields and countryside stretching out as far as the eye can see into the depths of the spheres, as well as more detailed proposals for mixed farm complexes that include extensive fields of grains, pulses, and vegetables.[16]

Nonetheless, the meat focus noticeable in the planning for the early space missions remained apparent in the overall planning of the food supply for these putative space colonies of the 1980s. One of the cited benefits of growing vegetables and grain hydroponically (where plants are grown through holes in Styrofoam boards, roots suspended in a nutrient-rich water solution) was that 'the roots could be harvested for animal feed.'[17] Although plant growing was seen as an essential component of the diet, meat, fish, and eggs were all assumed to be required for the health and, in particular, the contentment of the colonists: 'The colonists will want to eat meat.'[18] The focus was on being as efficient about it as possible. Rabbits, goats (for milk as well as meat), and chickens (for eggs as well as meat) were all proposed as ideal food animals, and an argument for the possibility of supporting a small herd of Hereford or other beef cattle was even put forward.[19] Although it was acknowledged that 'cattle are rather wasteful at converting feed to beef,' the overriding concern was that 'people like to eat beef.'[20] Growing plants for animal consumption and managing the additional smells and CO_2 build-up appear to have been considered prices worth paying to ensure as little change as possible to a meat-focused Western diet.

The Role of Vegetables in Conquering Space

The change in cultural expectations in the intervening decades is notable. Since the later 1980s and the introduction of the space shuttle, more fresh salad vegetables have been included in the food provisions, made possible by a small fridge and the use of irradiation to give longer life to fresh food. Changing diet expectations are also naturally reflected in the food plans for future space travel, in particular the ongoing investigation into the possibility of a manned trip to Mars. A mission to Mars would take at least two years, and could not be re-supplied en route (in contrast to the now-retired Mir and the current International Space Station [ISS], both in accessible orbit above Earth). Neither is it practical to provide room for the bulk and weight of two years' worth of provisions for a probable crew of six. Self-sufficiency is therefore required. The crew must be able to grow, process, cook, and consume food during both the outward and return journeys, and when they land on Mars. This requirement automatically gives hydroponically grown vegetables a starring role. ESA, the European Space Agency, expects 40 per cent of the diet to be grown and processed on board, the main crops being rice, onions, tomatoes, soya, potatoes, lettuce, spinach, wheat, and spirulina.[21] Indeed, meat has been dropped as a practical consideration, at least as a live commodity that could practically be supported and farmed en route and on arrival. The earth-bound debates on the energy required to raise meat for human consumption take on a new urgency and come to a speedier and more obvious conclusion when one is actually faced with the possibility of a crew of humans in confined conditions having to compete with their own dinner ingredients for the basic resources – water, air, food – needed for all of their survival.

Since the beginning of manned space programmes the conditions aboard spacecraft have made it almost impossible to replicate the food and eating experiences that the men and women involved would have on earth. The approach taken has been limited as much by cultural factors as by technical constraints. Historically, vegetables in space have reflected the cultural norms in US and Russian society; as side dish and minor ingredient in the former and soup ingredient and raw garnish in the other. Initially constrained by technology that processed them into often inedible and indigestible forms, vegetables are now poised to take a leading role in the future of space eating. Perhaps a love of vegetables will eventually become a new qualification for astronaut selection, at least for long-term missions.

Notes

1. Gagarin and Lebedev, 27.
2. Kanas and Manzey, 17.
3. In 1978 the USSR conducted the Opros questionnaire on board Salyut to help psychologists correlate physical and mental health; sense of smell was considered an important enough factor in the psychological condition of cosmonauts to be included with analysis of eating and sleeping habits. See Harland, 65; Rappole, 17.
4. See Compton and Benson; Lane and Schoeller; Sauer, ed.

5. Crouch, 58.
6. Harland, 343.
7. Harland, 42.
8. Harland, 150.
9. Harland, 101.
10. Sauer, ed., 20.
11. Hollender, 9.
12. Hollender, Klicka, and Smith, 272, 275.
13. Crouch, 246.
14. Crouch, 244.
15. Lineger, 201.
16. Web searches for Bernal or Dyson spheres readily bring up some of these illustrations. See also Heppenheimer; O'Leary.
17. Heppenheimer, 126.
18. Heppenheimer, 126.
19. Heppenheimer, 126–128.
20. Heppeneheimer, 128.
21. 'Dinner on Mars: How to cook Martian bread', at http://www.physorg.com/news4495.html.

Bibliography

Compton, W. David, and Charles D. Benson. *Living and Working in Space: A History of Skylab*. Washington DC: NASA, 1983

Crouch, Tom D. *Aiming for the Stars: The Dreamers and Doers of the Space Age*. Washington DC: Smithsonian Institution Press, 1999.

'Dinner on Mars: How to cook Martian bread' at http://www.physorg.com/news4495.html.

Gagarin, Yuri and Vladimir Lebedev. *Psychology and Space*, trans Boris Belitsky. Moscow: Mir Publishers, 1970.

Harland, David M. *The Mir Space Station: A Precursor to Space Colonisation*. Chichester, West Sussex: John Wiley & Sons Ltd., 1997

Heppenheimer, T. A. *Colonies in Space*. Harrisburg PA: Stackpole Books, 1977.

Hollender, Herbert A. 'Introductory Remarks' in *Activities Report: Feeding Systems in the Space Age*, Proceedings of the 24th Annual Meeting on Feeding Systems in the Space Age, April 14–16, 1970, Chicago, Illinois: Research & Development Associates for Military Food & Packaging Systems, Inc., 1970.

Hollender, Herbert A., Mary V. Klicka and Malcolm C. Smith. 'Food Technology Problems Related to Space Feeding.' pp. 265–279 in W. Vishniac & F.G. Favorite (Eds), *COSPAR Life Sciences and Space Research VIII – Proceedings of The Symposium on Nutrition of Man in Space Prague, 14 May 1969*. Amsterdam: North-Holland Publishing Company, 1970.

Kanas, Nick and Dietrich Manzey. *Space Psychology and Psychiatry*. El Segundo, California and Dordrecht, The Netherlands: Microcosm Press and Kluwer Academic Publishers, 2003.

Lane, Helen W. and Dale A. Schoeller. *Nutrition in Spaceflight and Weightlessness Models*. Boca Raton, Florida: CRC Press, 2000

Lineger, Jerry M. *Letters from Mir: An Astronaut's Letters to His Son*. New York: McGraw-Hill, 2003.

O'Leary, Brian. *The Fertile Stars*. New York: Everest House Publishers, 1981.

Rappole, Clinton L. 'Food Service Management' in *Food Service and Nutrition for the Space Station*, ed. Richard L. Sauer. Washington DC: NASA, 1985.

Sauer, Richard L., ed. *Food Service and Nutrition for the Space Station*. Washington DC: NASA, 1985.

Vishniac, W. & F.G. Favorite (Eds), *COSPAR Life Sciences and Space Research VIII – Proceedings of The Symposium on Nutrition of Man in Space Prague, 14 May 1969*. Amsterdam: North-Holland Publishing Company, 1970.

The History of the Potato in Irish Cuisine and Culture

Máirtín Mac Con Iomaire and Pádraic Óg Gallagher

Introduction
Few plants have been as central to the destiny of a nation as the potato (*Solanum tuberosum*) has been to Ireland. Ireland was the first European country to accept the potato as a serious food crop. From its introduction from South America, the potato has held a central place in the Irish diet, and by extension, in the culture of Ireland (Choiseul, Doherty et al. 2008:3). Potato growing is well suited to the Irish climate and soils, although both excessive and insufficient rainfall at certain times of the growing season can pose disease risks, the biggest of which is potato blight *Phytophthora infestans* (Lafferty, Commins et al. 1999:77). The potato's influence is to be seen in diverse spheres ranging from place names (Ballyporeen – the town of the little potato), folklore, literature, and poetry to the paintings of Paul Henry.

Ireland before the potato
The native Irish diet of cereal- and milk-based products augmented with pig meat survived relatively unchanged from prehistoric times to the introduction of the potato, possibly in the late sixteenth century (Sexton 2005:232). By the fourteenth century there was a fusion of Gaelic Irish and Anglo-Norman food patterns. Cullen (1992:47) suggests that Irish diets prior to the introduction of the potato were retarded, reflecting a medieval backwardness rather than poverty in the modern sense. The per capita consumption of butter in Ireland was the highest in the world. Meat consumption per capita was also relatively high and the range of meats eaten was uniquely wide, making the Irish diet and cooking, although relatively simple compared to the French, 'one of the most interesting culinary traditions in Europe' (Cullen 1981:141). Lucas (1960:8–43) provides a detailed account of food eaten before the arrival of the potato.

Arrival and assimilation
The potato was introduced to Europe from South America. Whether the introduction of the potato to Ireland can be credited to a Drake, Raleigh, or Southwell figure, or that they may have been washed ashore from wrecks of the Spanish Armada in 1588, it is clear that the potato had reached Ireland by the end of the sixteenth century (Sexton 1998:71; Salaman 2000:142–158). The potato transformed Ireland from an underpopulated island of 1 million in the 1590s to 8.2 million in 1840, making it the most densely populated country in Europe. Bourke (1993) mentions four phases of acceptance of the potato into the general Irish diet. Stage one (1590–1675) sees the potato used as a

supplementary food and standby against famine; stage two (1675–1750) sees the potato is viewed as a valuable winter food for the poorer classes; stage three (1750–1810) sees the poorer classes become dangerously reliant on potato as staple for most of the year; stage four (1810–1845) sees mounting distress as localized famines and potato failures become commonplace.

The potato was enjoyed by rich and poor alike, and Cullen (1992:46) points out that potatoes were exported from Ireland to the colonies and also suggests that Irish brandy merchants who settled in Cognac may have been the first to plant potatoes in the Charente region of France. Two centuries of genetic evolution resulted in yields growing from two tons per acre in 1670 to ten tons per acre in 1800 (Mac Con Iomaire 2003:209). Lyons (1982:35) notes that the potato was useful for cleaning, restoring, and reclaiming the soil, and also for fattening pigs. This point is elaborated by Cullen (1992:47), who suggests that increased potato consumption may simply and paradoxically reflect the fact that cereal cultivation intensified in the 1750s and 1760s, resulting in a growing reliance on the potato as a cleaning, restoring root crop. The potato provided the growing labour force needed for the move from pasture to tillage that occurred at this time, but resulted in high levels of unemployment following the Battle of Waterloo when the demand for exports fell.

The Irish had a peculiar way of cooking potatoes 'with and without the bone or the moon' (Wilde 1854:131). This method pertained to parboiling the potato leaving the core undercooked and was the preferred meal for a labourer with a day's work to do. The partially cooked potato lay in the stomach creating a second digestion period after the initial floury mass was digested that assisted in staving off hunger for longer periods for the worker. This is similar practice to the low Glycemic Index (G.I.) diet that athletes use today based on brown or *al dente* pasta, which releases energy gradually in the body. An urban-rural divide was prevalent in the art of cooking and eating potatoes. Wilde (1854:130) suggests that the cottier's-cabin potato 'wanted the flavour, the richness, the dryness, the fresh country look, and the dimple, the smile just bursting into a laugh,' and although rarely fully cooked, was superior to the town potato that 'had a sickly, cover dish flavour, and a would-be aristocratic air, which to those who knew better was quite disgusting.' Worse still, the town potato was peeled with a knife as opposed to that well groomed extra long thumb nail of the cottier who would 'be eating one potato, peeling a second, have a third in his fist and an eye on the fourth' (Sexton 1998:74). The Irish have always favoured floury varieties of potatoes to waxy varieties. This author's (MMCI) grandfather liked a potato that smiled at him – referring to a slight crack in the skin – not one that burst its sides laughing at him.

Not all Irish diets during the eighteenth and nineteenth centuries were dull and centred on the potato, dairy produce, and occasional bacon or pickled herrings. Although Ireland was the first European country to adopt the potato as a staple crop – a practice that spread to the colonies and to mainland Europe – European fashions in food and beverages also percolated Irish culinary practice.

The History of the Potato in Irish Cuisine and Culture

The potato in the Irish diet – class differences

Diet varied considerably with social status, the basic peasant staples of oats and dairy produce co-existed with the acquired traditions of the gentry. The introduction of the potato and other New World foods led to the narrowing of the diet of the poor and a broadening of the diet of the rich over the course of the eighteenth century. The Anglo-Irish ascendancy adopted some of the 'extraordinary hospitality' that had been part of the Gaelic tradition, but the conspicuous consumption was more sophisticated, emulating eating patterns in London and Paris. By the nineteenth century the potato was established as a staple of one-third of the population, an overdependence that led to the devastation of the Famine in the 1840s when successive harvests failed. The custom of preparing potato puddings, both sweet and savoury, was particularly noticeable among the wealthy, where extra ingredients like saffron, sugar, and spices differentiated this potato dish from the plain boiled potatoes of the cottiers (Sexton 1998:79). An article in *The Lady of the House* (May 1909) discusses antique potato rings that became fashionable in Georgian Ireland from around 1760–1790. The potato ring was made of silver and the bowl of potatoes, either wooden or other, would be sat on it. An inventory of Lord Viscount Doneraille's home in Kildare Street, Dublin in 1762 includes a tin potato roaster among such items as pewter plates, knives, and forks, tin oven for beefsteaks, coffee pot, gravy dish and cover, flesh forks, salt boxes, marble mortar, and wooden pestle (Griffin 1997:23–3).

Potatoes played a prominent role in public dining also. In 1823 a sign above one tent at Donnybrook Fair read 'sirloins, ribs, rounds, flanks, shins, brisket, six dozen boiled chickens, 28 Wicklow hams, kishes of potatoes, carts of bread, and gallons of punch' (Ó Maitiú 1996:22). Three tons of potatoes were served at the great Crimean Banquet, held in 1856, in a Dublin warehouse to honour the troops stationed in Ireland who had served in the Crimea, where 3,628 invited guests attended (Meredith 1997:57). The sight of four large vans of freshly cooked potatoes arriving enveloped with clouds of steam was reported in the newspapers the following day.

The misery of much of the Irish tenant farmers prior to the Famine, who having paid their rent had barely enough money or food to survive, is recorded in a conversation the French nobleman, Alexis de Tocqueville had with Thomas Kelly, Secretary of the Board of Commissioners of National Education, in 1835, when he posed the question:

> 'According to what you tell me, although the agricultural population is poor, the land produces a great deal?'

And the answer he received was:

> 'The yields are immense. There is no country where the price of farms is higher. But none of this wealth remains in the hands of the people. The Irishman raises beautiful crops, carries his harvest to the nearest port, puts it on board an English

vessel, and returns home to subsist on potatoes. He rears cattle, sends them to the London, and never eats meat.'

(Larkin 1990:29)

Pre-Famine potato varieties

An Irish writer, Rye (1730), in his work *Considerations on Agriculture* was the first to describe the different potato varieties (Choiseul, Doherty et al. 2008:4). Few of the pre-Famine potato varieties exist today. These varieties included the *Black* (pre-1730); the *Yellow*; the *Cluster;* the *Irish Apple* (1768); the *Red Nose Kidney* (syn. *Wicklow Banger*); *Cork Red*; the *Lumper* (pre 1808) and *Cup*. The period 1810–1845 saw the adoption of new inferior varieties of potatoes, notably the *Lumper*, which promised excellent yields. However, the strain was not resistant to the potato blight, and this resulted in the dramatic potato failures of 1845, '46 and '47.

The Great Famine

The potato failures of 1845–1847 resulted in devastating famine which left the country socially and emotionally scarred for well over a century. Descriptions of the appalling conditions of the poor cottier class during the famine years, along with economic analyses, are found in the writings of Woodham-Smith (1991) and Ó Gráda (2000). The Great Famine affected the poorer cottier class who had developed an unhealthy dependence on the potato in their diet. Sustaining a family solely on a diet of potatoes required consuming vast quantities of the tuber. Bourke (1968:73–78) attempts to define the average daily amount and settles on 12 pounds per adult male (one over 15 years as defined by the 1841 census) as this takes into account spring scarcity of the crop. This monotonous diet was augmented by butter, buttermilk, salt, onions, occasional bacon, and salted herrings. The result of the Great Famine was that by 1851, at least one million of the Irish poor had died and another million had emigrated (Sexton 1998:74). The inescapable link in folk memory between the Famine and the potato is illustrated in Patrick Kavanagh's poem 'Restaurant Reverie':

> O half potato on my plate,
> It is too soon to celebrate
> The centenary of '48
>   Or even '47.
> You're boasted in the centre, too,
> And wet, in soapy soil you grew,
> But I am thankful still to you
>   For hints of history given.
>
> There's something lonely far away
> In what you symbolise to-day

> For me – the half that went astray
>   Of life, the uncompleted.
> But up brown drills new pink buds start
> With truer truth than truth of art,
> Ignoring last crop's broken heart
> And a generation defeated.
>   Oh, here is life
>   Without a wife
> [A] half-potato. Eat it!
>   (Kavanagh 1972:137)

Bourke (1959) calculated that 829,875 hectares were under potato cultivation in 1845. He also calculated a pre-Famine potato balance sheet from which he estimated that the three principal uses of the 15 million tons of potatoes were as human food (47 per cent), animal feed (33 per cent), and seed (13 per cent), with a further 2 per cent allocated for export and the remaining 5 per cent attributed to loss or waste (C.S.O. 1997:18). The middle and upper classes, however, were relatively unaffected by famine and retained a varied diet before, during, and after the Famine that would be 'hard to surpass in contemporary rural France or Britain' (Cullen 1981:162–3). The farming class also escaped the worst effects of the Famine. For example, the diet of both family and fed-labourers of a 30-acre north Dublin farm, on the eve of the Famine, seemed rich:

> The food was nearly all home made: wholemeal bread; oaten meal grown on the farm made into stirabout; potatoes, generally all floury; first quality butter; bacon, raised, killed and cured on the premises; milk unadulterated '*ad libitum*' for everyone and everything and honey bees in almost every garden.
>   (Kettle 1958:5–6)

Post-Famine potato varieties

Despite the devastation of the Famine, it was not until 1880 that the *Report of the Select Committee of the House of Commons* decided that research and breeding of new blight-resistant varieties was of national importance (Dowley and O'Connor 2007:7). Thomas Carroll of the Albert College, Dublin, with the help of farm superintendents of the Department of the National Board of Education, organized a crossing programme, producing true seed that was distributed to schools with instructions for growing and selection that resulted in 255 new varieties being exhibited at the Royal Dublin Society's 1885 winter show and ultimately led to the popularizing of the more blight-resistant varieties such as the *Champion, Skerry Champion,* and the *Shamrock*. In 1847, the first year potato acreage was recorded, the *Rock* variety accounted for 40 per cent of the crop and maintained a dominant position until the *Champion* proved remarkable resistance to the *Phytophthora infestans* outbreak of 1879. Within four years the *Champion* became the dominant potato accounting for 80 per cent of the total crop and remained popular until the 1900s when its resistance to blight began to diminish (Choiseul, Doherty et al. 2008:5–7).

Potato breeding declined in Ireland during the first half of the twentieth century with UK varieties *Kerr's Pink, Arran Banner, Majestic,* and *King Edward* and the Dutch variety *Record* replacing the *Champion* (Dowley and O'Connor 2007:7). Potato breeding didn't recommence in Ireland until the middle of the twentieth century when the Department of Agriculture started a breeding programme in Donegal that culminated in the release of the first early variety, the *Irish Peace*. The potato breeding programme in Oak Park, Co. Carlow commenced in the 1960s and at first was primarily concerned with producing a blight-resistant replacement for *Kerr's Pink* for the domestic market (Dowley and O'Connor 2007:7).

Many of the varieties that have dominated the Irish market in recent history have their origins in late nineteenth- and early twentieth-century varieties such as the *British Queen* (1894), *King Edward VII* (1902), *Golden Wonder* (1906), *Kerr's Pink* (1907), and *Record* (1925). The fact that they have prevailed is a testament to their breeders and to the progress in seed production (Choiseul, Doherty et al. 2008:6). One of Oak Park's most recent successes has been the launch of the *Rooster* variety. This potato accounts for 45 per cent of the Irish market today while *Kerr's Pink* holds 25 per cent. The total area of potatoes grown in 1991 was just over 20,000 hectares, with the highest concentration of commercial production located in north Dublin, east Donegal, and parts of counties Meath and Louth. Seed-potato growing is confined primarily to east Donegal since the lower temperatures curtail the transmission of viral disease (Lafferty, Commins et al. 1999:77). In the last 30 years the total area of potatoes declined in all regions, but most significantly in the western counties, midlands, and mid-west. Between 1980 and 1997, the area under potatoes declined by 56 per cent, or some 23,400 hectares.

Competing carbohydrates

Today's principal competition to the potato among carbohydrates comes from bread, pasta, and rice. At the turn of the twentieth century, Italian immigrants engaged in promoting a new method of potato cookery rather than spreading the use of pasta. An article in the *Evening Mail* in October 1927, discussing the previous 30 years of Dublin life, writes of a strong Italian community in Chancery Lane who sold chipped potatoes from mobile cooking shops at various points in the city. These open-air chip shops along with the street coffee booths were extremely rare by the late 1920s (G.D. 1927). Ice-cream parlours and later fish and chip shops were opened during the twentieth century by Italian immigrants such as the Cinelli, Borza, Forte, Fusciardi, Fusco, Morelli, Tosselli, and Cervi families (Power 1988; Reynolds 1993). Fried potatoes were offered along with mashed potatoes and sauté or Lyonnaise potatoes on the à la carte menu of The Plaza Restaurant in Dublin in 1928. Boiled potatoes, potatoes with cream, *pommes fondants,* and *pommes parmentier* were also offered as part of the table d'hôte menu in The Plaza (Geldof 2003). Most restaurants in Ireland during the twentieth century served potatoes either plainly boiled or prepared in one of the many methods of the French culinary canon. The rise of convenience foods such as instant mashed potato in the 1970s led to the ridiculous situation of country hotels, surrounded by fields of beautiful Irish potatoes, serving instant mash out of a packet to an uncritical public. Some culinary leaders such as Myrtle Allen in Ballymaloe House and Dick Fletcher in The Galley Restaurant served 'kishes' of new Irish potatoes with pride.

Indian and Chinese restaurants began to appear in Dublin in the 1940s and 1950s offering both rice and noodles (Burke 1941; Mac Con Iomaire 2006a; Lee 2008). Farmar (1991:180–182) notes that taste in food among the Irish became more adventurous following the growth of international package holidays and increased airline travel in the 1960s and 1970s. Despite a growing interest in pasta, rice, and noodles, potatoes remained the dominant carbohydrate for Irish diners throughout the twentieth century.

Irish potato dishes

During the last 20 years there has been a trend to serve interpretations of traditional Irish dishes in up-market restaurants. Dishes such as champ, colcannon, bubble and squeak, boxty, potato cakes, clapshot, and potato-rich dishes like Irish stew and Dublin coddle have increased in popularity. Traditionally, July was always known as 'hungry July' as the old crop had finished and the population waited to the first day of August to pick the new crop. Sexton (1998:70) likens the excitement caused when Irish new potatoes go on sale in early summer to the arrival of the new year's Beaujolais crop in France. Nowadays in summertime, Wexford new potatoes and strawberries are sold from mobile units along main roads the length and breadth of Ireland. An unfortunate reality of trends in food retailing is that it is almost impossible to purchase new Irish potatoes in Irish supermarkets, where new potatoes from Cyprus, France, Italy, and Israel are far more available.

Boxty

Although the kitchen skills of the peasant were basic, Wilde (1854:131) claimed that every peasant girl could tell when a potato or egg was cooked just by holding it. Wilde further discusses Irish potato cuisine that he believes did not exist outside of Ireland and mentions two or three dishes not found in cookery books. He describes the making of boxty or boxtie, as it is popularly known in the west of Ireland (1854:133). Wilde claims boxty was known as scotchy, buck-bread, or stampy in the south of the country and was very popular with the children.

Wilde (1854:133) describes how the children would make a grater out of old tin cans by punching holes in the side of the can using an awl. A similar description of making a home-made grater is given by Uí Chomáin (1992:53), who recounts first seeing boxty made in her grandmother's home and describes how the home-made grater was made:

An empty tin-can, of the type that holds peas or beans, had both circular ends removed. The remainder was flattened out and holes bored all over it. This was nailed to a piece of wood larger than the tin, with the rough side of the tin facing upwards. The peeled raw potatoes were then grated into a basin.

(Uí Chomáin 1992:53)

There are three types of boxty: boiled boxty, also known as boxty dumplings, pan boxty, and loaf boxty (Irwin 1937; Allen 1995). Recipes differ from parish to parish but the main ingredients are equal amounts raw and cooked potatoes with the addition of varying amounts of flour (Uí Chomáin 1992:53). Sexton (1998:82) describes stampy as a deluxe version of boxty bread containing cream, sugar, and caraway seeds.

Boiled boxty is the most time-consuming to produce. It necessitates peeling and grating raw potatoes, squeezing excess liquid from the pulp, adding the pulp to an equal amount of mashed cooked potatoes, binding with some flour, seasoning with salt, kneading, and rolling into large balls about 12 centimetres in diameter. These are then placed in a pot of boiling salted water, returned to the boil and when the dumplings rise to the top let simmer for up to 45 minutes. The boiled boxty is then let cool. It is then sliced and refried in some butter in a pan.

For boxty loaf or baked boxty, proceed as for boiled boxty; after kneading, place in a greased loaf tin and bake in a moderate oven for one hour.

To make boxty pancakes or pan boxty, mix equal amounts of grated raw potatoes (with excess water squeezed out) and mashed cooked potatoes with an equal amount of flour seasoned with salt. Enough milk is then mixed in to make a batter of dropping consistency. This batter is then fried as pancakes in a pan over a medium heat, turning once. Some recipes call for the addition of egg, although the more traditional would not. Others call for a teaspoon of bread soda (Irwin 1937; Fitzgibbon 1983). As with boiled boxty, pan boxty tastes better if allowed to cool, and then reheated.

The key to making boxty is to remove excess liquid from the potato. First the raw potatoes are grated into bowl, then placed in a muslin cloth and the excess liquid squeezed out and gathered in another bowl. This liquid is allowed settle and the starch will separate and sit on the bottom of the bowl. The liquid can then be poured off leaving potato starch in the bottom of the bowl. Some recipes require the starch to be added back to the boxty mix (Allen 1995:152).

Champ and colcannon

Recipes for champ, a potato dish made with mashed potatoes and scallions softened in butter and milk, and colcannon, mashed potatoes with curly kale or green cabbage, were also collected by the National Folklore Commission. It is said that the Irish brought colcannon to England with them and that leftover colcannon that was fried up on the pan for breakfast became known as bubble and squeak. Another dish called Kala, popular in West Galway, was made from mashed potato, butter, chopped onions topped with a boiled egg (Mac Con Iomaire and Cully 2007:146). Another popular potato dish is potato cakes, which are made from mashed potatoes and flour, with some recipes adding a little baking powder to make them lighter. They are fried either dry or with butter on a pan and eaten with butter. Sexton (1998:78) notes that both boxty and potato cakes were portable and could be carried as a snack to ward off hunger.

Potato bread is not unique to Ireland and most countries where the potato was adopted have their own peculiar fashion of processing the humble tuber. *Lompe* and *Lefse* are popular in Norway, *Roesti* in Switzerland, *Reibekuechen* in Germany and *Kartoffelpuffer* in Bavaria, *Latkes* in Jewish cuisine, all of which are potato breads or pancakes.

Conclusions

Ireland was the first European country to adopt the potato as a serious food crop. From its introduction from South America the potato went through four stages of growth up until the devastation of the Famine (1845–1849). The varieties of potatoes eaten ranged over the centuries from the *Irish Apple, Cup, Lumper, Rock, Champion, Kerr's Pink*, to the *Rooster* which today holds the dominant market position. Potatoes were eaten at all levels of society, but in different guises. Irish people have traditionally preferred floury potatoes to waxy varieties. Whilst silversmiths in Georgian Ireland made potato rings for the Anglo-Irish ascendancy, the poor cottiers cooked in a cauldron and ate their potatoes 'with and without the moon,' using a long thumb nail to peel the skin. Potato production has declined dramatically in Ireland from 829,875 hectares in 1845 to 20,000 hectares in 1991. Despite the decline, we still look forward to new potatoes each summer as the French anticipate their Beaujolais. Traditional potato dishes such as colcannon, champ and potato cakes remain popular. Boxty seems to have been more popular in the north-western counties of Sligo, Leitrim, Roscommon and Longford than elsewhere. Boiled boxty could be considered a unique Irish product or, as one individual called it, 'the caviar of North Longford' (Gallagher 2008).

References

Allen, D. *Irish Traditional Cooking*. London: Butler & Tanner, 1995.
Bourke, A. *The Extent of the Potato Crop in Ireland at the time of the Famine*. Dublin: Statistical and Social Inquiry Society of Ireland, 1959.
Bourke, A. *The Use of the Potato Crop in Pre-Famine Ireland*. Dublin: Statistical and Social Inquiry of Ireland, 1968.
Bourke, A. *The Visitation of God? The Potato and the Great Irish Famine*. Dublin: Lilliput Press, 1993.
C.S.O. *Farming since the Famine: Irish Farm Statistics 1847–1996*. Dublin: Central Statistics Office, 1997.
Choiseul, J. W., G. Doherty et al. *Potato Varieties of Historical Interest in Ireland*. Dublin: DAFF, 2008.

Cullen, L. M. *The Emergence of Modern Ireland 1600–1900*. London: Batsford Academic, 1981.
Cullen, L. M. Comparative Aspects of Irish Diet, 1550–1850. In *European Food History: A Research Review*, ed. H. J. Teutberg, 45–55. Leicester: Leicester University Press, 1992.
Dowley, L. J. and N. O'Connor. *The Oak Park Potato Varieties*. Carlow: Teagasc, 2007.
Farmar, T. *Ordinary Lives: Three Generations of Middle Class Experience 1907, 1932, 1963*. Dublin: Gill and Macmillan, 1991.
Fitzgibbon, T. *Irish Traditional Food*. Dublin: Gill and Macmillan, 1983.
G.D.. 'Thirty Years of Dublin Life: Changes and Events.' *Dublin Evening Mail, 1927*.
Gallagher, P. Ó. *Boxty – 'The Caviar of North Longford': An investigation into the Origins and Peculiarities of the Traditional Irish Potato Dish, Boxty* (Unpublished Thesis). School of Culinary Arts and Food Technology, Dublin Institute of Technology, 2008.
Geldof, H. Herbert (Sonny) Geldof Interview with Máirtín Mac Con Iomaire in Sadymount, Dublin (3/4/2003).
Griffin, D. J. 'The building and furnishing of a Dublin townhouse in the 18th century,' *Bulletin of the Irish Georgian Society*. XXXVIII – 1996–1997: 24–39.
Irwin, F. *Irish Country Recipes*. Belfast: The Northern Whig Ltd, 1937.
Kavanagh, P., ed. *Kavanagh – The Complete Poems*. Newbridge: Goldsmith Press, 1972.
Kettle, L. J., ed. *Material for Victory: The Memoirs of Andrew J. Kettle*. Dublin, Fallon, 1958.
Lafferty, S., P. Commins, et al. *Irish Agriculture in Transition: A Census Atlas of Agriculture in the Republic of Ireland*. Dublin: Teagasc, 1999.
Larkin, E., ed. *Alexis de Tocqueville's Journey in Ireland July-August 1835*. Dublin, Wolfhound,1990.
Lee, M.. Personal communication with Mary Lee – daughter of Hong Kong restaurateur and head chef of the Luna, O'Connell Street, Dublin in the 1950s (7/3/08).
Lucas, A. T. 1960. Irish Food before the Potato. *Gwerin* III(2):8–43, 1960.
Lyons, F. S. L *Ireland since the Famine*. London: Fontana, 1982.
Mac Con Iomaire, M.. 'The Pig in Irish Cuisine past and present' in *The Fat of the Land: Proceedings of the Oxford Symposium on Food and Cookery 2002*. ed. H.Walker. 207–215. Bristol: Footwork, 2003.
Mac Con Iomaire, M. 'Mike Butt.' In *Culinary Biographies*. ed. A. Arndt. 85–86. Texas: Yes Press, 2006.
Mac Con Iomaire, M. and A. Cully. 'A History of Eggs in Irish Cuisine and Culture,' in *Eggs in Cookery: Proceedings of the Oxford Symposium on Food and Cookery 2006*. ed. R. Hosking. 137–149. Totnes, Devon: Prospect Books, 2007.
Meredith, J. *Around and about the Custom House*. Dublin: Four Courts Press, 1997.
Ó Gráda, C. *Black '47 and beyond: The Great Irish Famine in history, economy and memory*. Princeton, New Jersey: Princeton University Press, 2000.
Ó Maitiú, S. 'Donnybrook Fair: carnival versus lent,' *History Ireland* 4(1): 21–26, 1996.
Power, U. *Terra Straniera: The Story of the Italians in Ireland*. Dublin, Club Italiano, 1988.
Reynolds, B. *Casalattico and the Italian community in Ireland*. Dublin: UCD Foundation for Italian Studies, 1993.
Salaman, R. *The Social History of the Potato* (revised impression edited with a new introduction by J. G. Hawkes). Cambridge: Cambridge University Press, 2000.
Sexton, R. *A Little History of Irish Food*. Dublin: Gill and Macmillan Ltd, 1998.
Sexton, R. 'Ireland: Simplicity and integration, continuity and change,' in *Culinary Cultures of Europe: Identity, diversity and dialogue*. ed. D. Goldstein and K. Merkle. 227–240. Strasbourg: Council of Europe Publishing, 2005.
Uí Chomáin, M. *Cuisine le Máirín*. Dublin: Attic Press, 1992.
Wilde, W. 'The Food of the Irish.' *The Dublin University Magazine* XLIII(CCLIV):127–146, 1854.
Woodham-Smith, C. *The Great Hunger*. London: Penguin, 1991.

'Sweet as'–
Notes on the Kumara or New Zealand Sweet Potato as a *Taonga* or Treasure

Ray McVinnie

Like many Aucklanders, the people of New Zealand's largest city, I live on the slopes of one the region's many extinct (we hope) volcanoes, known as 'Maungawhau.' Anyone who walks over these grassy cones can tell you that European New Zealanders like myself were not the first people to live here. Today Maungawhau still shows clear signs of the terraced contours and ditches that made up the defences of a traditional Maori *pa* or hill fort, which the area's first inhabitants either retired to in times of war or permanently inhabited. Also evident are numerous shallow pits of great significance because many were used to store kumara – sweet potato (*Ipomoea batatas*). This tuber was the Maoris' most important food crop. Evidence of pre-European kumara gardening and storage is found over most of the North Island and the top half of the South Island of New Zealand. This paper will give a brief account of how the kumara came to be considered a *taonga*, or treasure, by both Maori and *Pakeha* (European) New Zealanders.

Several conflicting scientific theories attempt to explain how the kumara came from its home in South America to eastern and then western Polynesia, but it was with the Maori that it later came south to New Zealand about 1000 years ago. Transporting this tropical plant by sea in open canoes and successfully cultivating it 1000 kilometres south of anywhere it had previously grown testify to the Maoris' skills as cultivators. They brought other plants from their ancestral homeland of Hawaiiki, such as taro, gourd, paper mulberry tree, breadfruit, tropical cabbage tree, and coconut. Most failed in New Zealand's colder climate. Of those that survived, the kumara, taro, and yam were mostly restricted to the warmer North Island. In contrast to tropical Polynesia, agriculture was difficult even in subtropical regions. Kumara emerged as the pre-eminent food crop with the most southerly extent of cultivation. This was because it '…freely responded to care and attention in the most varied situations and yielded a large crop of an article at once palatable, wholesome, and nutritious. With the primitive Maori, in fact, the kumara stood in a class by itself, above and apart from everything else. As the mainstay of life it was regarded with the greatest respect and veneration. It was celebrated in song, and story, and proverb. Its cultivation and treatment called forth the utmost care and ingenuity, and were accompanied by the strictest and most elaborate religious observances.'[1]

Early European accounts demonstrate admiration for Maori agriculture: 'The whole *tout ensemble* was really admirable! The extreme regularity of their planting, the kumara… being generally set about two feet apart, in true quincunx [offset spacing] order, with no

Notes on the Kumara

deviation from a straight line when viewed in any direction...; the total absence of weeds, the care in which all was kept.'[2] The great skill of this stone-age people in modifying and developing their tropical horticultural practices ensured the kumara's success.

The earliest known pre-European kumara gardening sites date back to AD 1300. The two largest were in Marlborough at the top of the South Island and in Wairarapa at the bottom of the North Island, each 2000 hectares in area.[3] Evidence for pre-European horticulture consists of remains of specialized or unusual methods of cultivation. Many areas probably needed little modification, and have consequently disappeared.[4] Maori compensated for profound changes in climate and soil quality, selecting kumara-growing sites in microclimates, often on north-facing coastal slopes.[5] Remains of stone structures are still visible – rows, alignments, mounds and heaps possibly used as windbreaks, retaining terraces and boundary markers, or merely to tidy away stones cleared from fields. There is also evidence of drainage ditches and channels. Excavation pits remain where gravel and sand were removed to alter the composition of garden soil. Some gardens show evidence of soils being improved by such additions.[6] Increasing gritty content was a method of producing free-draining soil that trapped and held the sun's heat, essential for frost-tender, heat-loving plants like the kumara.[7]

In tropical Polynesia kumara was grown as a perennial from cuttings; in colder New Zealand it was treated as annual with the planting of sprouted tubers.[8] Unique, effective methods for storing kumara food and seed over winter were also developed – semi-subterranean storage pits, still discernable on *pa* sites. Some *pa* undoubtedly existed to defend food stores, illustrating the value of kumara.[9]

Maori agriculture was also intriguing in '...their national non-usage of all and every kind of manure; unless, indeed, their fresh annual layers of dry gravel...may be classed under this head. But their whole inner-man revolted at such a thing; and when the first missionaries first used such substances in their kitchen gardens it was brought against them as a charge of high opprobrium...They also never watered their plants, not even in times of great drought, with their plantations close to a river, when doing so might have saved their crops.'[10] The explanation for this may lie in the kumara fields' ritual status while plants were growing, which meant they were '*tapu*.' This word has strong connotations of sacredness. Only those ritually designated could enter fields, so manure by its very nature and water from elsewhere may have been regarded as defiling substances.

Kumara was significant both as a highly nutritious food and a spiritual and physical link to Hawaiiki.[11] Only on the east Polynesian island of Rapa Nui was its status as high as in New Zealand. Such esteem is evident in rituals associated with growing kumara and in the oral traditions describing canoe journeys back to Hawaiiki, undertaken specifically to bring the kumara to New Zealand.[12] Maori brought back many varieties differing in size, shape, colour, and function, known for special qualities such as sweetness, flavour, or yield.[13] In 1880 Colenso could list fourteen white-fleshed white-skinned varieties, three white-skinned reddish-fleshed varieties, seven red-skinned and fleshed varieties and eight dark purple-skinned and fleshed varieties, all of which came from the north of New Zealand. He lists another sixteen varieties from the eastern North Island and comments further, 'I do not consider the foregoing lists as being anything like exhaustive (indeed I have the names of a few others from the north which I purposely keep back); many of them I have both seen and eaten, 40 years ago and more.'[14]

Several varieties were already known to be lost at the time.[15] Extinctions continued as *Pakeha* introduced the more tolerant potato. Eventually kumara cultivation was restricted to a few larger, sweeter varieties. One of the most popular was the '*merikana*' introduced by American whalers.[16] Today there are three varieties grown commercially, 'Owairaka Red,' cream-fleshed, purple-skinned and considered a traditional cultivar, the 'Toka Toka Gold,' creamy-yellow-skinned and yellow-fleshed, and the 'Beauregard,' orange-skinned and fleshed, introduced from the USA.[17] However, traditional white-skinned white-fleshed varieties were identified in Japan and brought back to New Zealand in 1988. They are still being researched to see how they survive transplanting, to determine pest resistance, and to gauge reaction to climatic differences.

Kumara cultivation was an elaborate and necessarily highly cooperative undertaking. Choice of site for such an important crop was crucial, especially for smaller, weaker communities. With so much effort required, any incursion by hostile tribes could mean disaster for cultivators. Consequently, plantations were often scattered and located out of sight if difficult to defend.[18] Segregation of kumara plots was also a function of the rituals observed around its planting and harvesting.[19]

In October or early November the *pipiwharauroa* (shining cuckoo) returned from islands on the Pacific's western edge. Its characteristic call – '*koia koia*' – translated as 'dig, dig' or 'dig away,' signalled ground-preparation time for planting.[20] The stars also marked stages of kumara cultivation. In addition, a 'mackerel sky' around October resembling a cultivated kumara field indicated that the gods were preparing a celestial kumara plot and humans should be doing the same on earth.[21] After burning vegetation to clear the land, roots were removed from the soil and the drainage channels were dug where necessary. Conveniently, this was also the time when the edible fern root was at its best.

'In those plantations all worked alike: the chief, the lady and the slave; and all while so engaged, were under a rigid law of minute ceremonial restrictions, or taboo, which were invariably observed...It was a pretty sight to see a chief and his followers at work in preparing the ground for the planting of the kumara. They worked together, naked, (save for a small mat or fragment of one about their loins), in a regular line or band, each armed with a long handled spade (*koo*) [ko], and like ourselves in performing spade labour, often enlivening their labour with a suitable chaunt or song, in the chorus of which they all joined in.'[22] The *ko* was an elaborately carved long wooden stick with a sharpened flattened end like a narrow spade, with a footrest lashed to it. Its end was sometimes carved in the shape of a head and decorated with feathers.[23] The idyllic description above tends to belie the seriousness of the process. Preparing the ground required the whole community, the *ko*-ing done by men with the women and children following to break up large lumps of earth with smaller wooden tools.

Notes on the Kumara

The *tohunga* or priestly adept attended all stages of cultivation to deliver prayers and chants necessary for a good crop. Digging occurred in choreographed lines and chants ceremonially invoked the gods to bless the labour.[24] Plantations could be hundreds or even thousands of hectares in area, so annually the total effort expended was clearly staggering, and the social cohesion needed for a successful harvest was reflected in the crop's health and abundance. A divided community lacked the resources needed to achieve these ideals, no matter how favourable the conditions.[25] Customs varied from tribe to tribe, but from planting to storage kumara plantations themselves and those working on them were considered *tapu* (consecrated or sacred). Breaching *tapu* or even approaching plots too closely could easily invoke the death penalty. Even those admitted must take care, only entering from the north; any breaks in southerly, least favourable plantation boundaries could expose the crop to cold winds. People entering from east or west might cast shadows that could ruin the crop.[26] Such social controls protected this important crop from outsiders, ensuring it was raised according to strict methods crucial for its survival. Planting involved the chiefly or *rangatira* class of both sexes, but the presence of women in the kumara fields varied depending on local custom. There were often stone tutelary images present to which offerings were made.[27] Exhumed bones of ancestors and preserved heads of slain enemies were considered powerful talismans, often used at planting or when crops were not thriving.[28]

In pre-European times constant weeding and constructing fencing apart from windbreaks were largely unnecessary, as many weeds and animal pests only arrived with *Pakeha*.[29] The most significant threat was the *hotete*, the five-centimetre-long caterpillar of a large moth that could appear in great numbers and strip kumara plants of their leaves. These were picked off by hand. Colenso recorded an enterprising chief borrowing his turkeys to clear caterpillars from his plot. Whether or not turkeys breached *tapu* was an interesting theological point debated at the time.[30]

March or April was harvest time. Kumara were lifted for storage, requiring very close attention as dry storage at the right temperature avoided food and seed kumara rotting over winter. Storage solutions included roofed pits lined with crumbled dry wood and gravel, and carefully designed buildings above ground which were sometimes richly carved, stained red, and decorated with iridescent *paua* (New Zealand abalone) shells.[31] Both structures were strictly *tapu*.

Kumara cultivation required effort by much of the community almost year-round, but the tuber was only available as food for part of the year. Growing the crop had social significance that certainly equalled and probably exceeded its value as an important food source. Supervision of the yearly round of cultivation by leaders strengthened their status within the tribe, and the reputations of individual chiefs and whole tribes were greatly enhanced by the ability to provide the prized kumara for hospitality, trade and ceremonial feasts. It was this aspect of the kumara that ensured continued cultivation of a temperamental plant in hostile conditions.[32]

The kumara's cultural significance is illustrated by its central place at great feasts, or *hakari* held at harvest and on *rites de passages* such as births, deaths, betrothals, marriages, the exhumation of bones, or the building of a chief's meeting house. To qualify as a luxury item worthy of being used in these ceremonial food distributions or potlatches, this hard-earned commodity was amassed in huge quantities.[33]

The amount of food at these feasts was gargantuan, as Colenso remarks: 'At a small feast (comparatively) of this kind, …held at Waimate (Bay of Islands) in 1835, and given to the people of Hokianga, 2000 one-bushel baskets of kumara were used; and at a similar feast given by the noted warrior Te Waharoa…at Matamata, in 1837, to the people of Tauranga, the following inventory of the food was taken down by a credible eye-witness: "Upwards of 20,000 dried eels, several tons of sea fish, principally young sharks…a large quantity of hogs, 19 big calabashes of shark oil, 6 albatrosses and baskets of potatoes (sweet and common) without number."'[34] The food was displayed on giant cone-shaped, tiered, wooden structures called *potehe* which could be up to 30 metres high with a 10 metre circular base: '…when filled, they present one solid mass of food; the whole is decorated with flags, and, when in an elevated situation, presents a very imposing appearance. The portion belonging to each tribe is particularly pointed out: and when the ceremony of presenting is over, the people carry away their portions…[35]

For Maori everything has a *whakapapa*, including the kumara. 'When applied to humans, this word refers to genealogies or family trees, with the implications of shared genetic relationships and descent from a common ancestor among all persons named in that *whakapapa*. But when applied to non-human things (e.g. plants, animals, rocks, and stars), it is clear that other factors such as habitat and morphology … provide an important rationale for each grouping.'[36] The function of *whakapapa* and its accompanying allegorical narrative is concerned with advice about correct behaviour in dangerous or unpredictable situations, giving order to a complicated environment, assisting understanding of ethical questions and reinforcing the value of important cultural mores.[37] All Maori have a *whakapapa*, a thousand-year-old genealogy to be recited tracing all pedigrees back to one of the renowned captains of the ancestral canoe fleet from Hawaiiki. These men were the descendents of humans engendered by the gods, themselves the children of the two creation story protagonists, Ranginui, the sky father, and Papatuanuku, the earth mother.[38] Though all *whakapapa* identify a common ancestor, human *whakapapa* are true genealogies because they concern one species with shared inherited genetic characteristics.[39] Another traditional Maori concept inextricably linked with *whakapapa* is '*mauri*,' translatable as a life force or that which joins the physical and spiritual aspects of a person. It has been defined as the spiritual side of *whakapapa*.[40]

According to the kumara's *whakapapa* one of Ranginui and Papatuanuku's 'god children' is Rongo, considered in all the tribal variations of the *whakapapa* as the god of peace and cultivated foods, including kumara. The children of Rongo and his wife Pani represent the different kumara varieties possessed by the Maoris. Related to Rongo are Whanui, (the star Vega, whose appearance also meant it was time to harvest kumara), several unrelated plants with similar vine-like foliage, the unwelcome *kiore* or native rat which ate the kumara, and the caterpillars which plagued the kumara plants. This 'story' and the part played by each character in it is the *whakapapa* of the kumara. It is an allegorical body of knowledge containing every important aspect of kumara cultivation. Unlike a human *whakapapa*, this one contains information about insects, animals, celestial bodies, other plants and the earth itself.[41]

Clearly, human and non-human *whakapapa* are distinct entities, a vital consideration in debate about genetic modification of organisms. This particular way of ordering the world, granting everything its own genealogy, an accompanying narrative, and a life force connected with sacredness, authority, and integrity, is inherent in Maori life and culture. Its legitimacy is beyond question.

Inevitably, reconciling such a view of life with the strand of science that promotes the transferring of genes from one species to another through human intervention presents difficulties. Self-evidently, Maori would regard this as a transgression and a corruption of *whakapapa*.[42] In addition there are practical proprietorial questions raised by the intervention of modern scientific endeavour in such areas of cultural sensitivity. Like other *taonga*, kumara is guaranteed political protection by the Treaty of Waitangi, the founding agreement that established the relationship between Maori and the British Crown.[43] Awareness of traditional Maori beliefs is increasing among *Pakeha*, and the political implications of genetic modification on *taonga* are of concern to all New Zealanders. Maori debate on genetic modification of organisms always occurs in the context of Maori traditional beliefs. Do such issues as the fact that the kumara's *whakapapa* includes different species mean it is acceptable to transfer genes between species, or does each genome possess a *mauri* that should not be tampered with?[44] Any resolution depends on understanding concepts such as *whakapapa* and an appreciation of Maori as '…a people who walk backwards into the future, a reference to the importance placed on seeking guidance for future actions from the wisdom of the past deeds of ancestors and mythical heroes.'[45]

Maori cooked food in *hangi* or earth ovens as they had no ceramic tradition. Some tribes living near thermal springs boiled or steamed food in flax baskets suspended over or in hot pools. The *hangi* was made by heating stones to a very high temperature in a shallow trench with fire, raking out the fire, adding a little water to blow away the ashes, laying the food in flax baskets on top, covering it with fern fronds, flax mats, and finally earth, and waiting for about an hour for the food to cook.[46] This method slow-cooked the kumara, causing it to sweeten due to the effect of heat on an enzyme that converts starch to maltose. Some moist varieties become so sweet they taste as though they have been dipped in syrup. The process of converting starch to maltose starts at 57°C and finishes at 75°C, at which point the enzyme loses its effectiveness. Fast cooking at high temperatures will not achieve this effect.[47] According to the New Zealand Fresh Vegetable Industry website, kumara also gets sweeter the longer it is stored.

Notes on the Kumara

Maori had another highly prized form of kumara called *kao* kumara. Small varieties were chosen, washed, scraped with fern, and dried in the sun for two or three days, and then put into a large *hangi* for 24 hours. They were removed, dried again, and then stored in baskets in elevated storehouses or *patakas*. The *kao* kumara was dry and black with a very sweet aromatic flavour and was effectively preserved, lasting up to 2 years if kept dry. It was eaten crumbled and mixed with water and used as dry provisions on journeys. *Kao* was highly prized and large amounts (in one account 30 or 40 basketfuls) were prepared.[48]

With the arrival of *Pakeha* in the nineteenth century, Maori began to grow more easily raised crops like potatoes and maize. The reliance on kumara for food and as a ceremonial crop began to wane. By the time *Pakeha* arrived in large numbers most European vegetables were already being grown. There are early accounts of *Pakeha* eating indigenous New Zealand food plants but these were regarded as curiosities and disregarded as more European foods were imported.[49] The exception was kumara, the only indigenous food adopted by *Pakeha* and still eaten by this group today. The explanation, apart from kumara's appealing sweet taste, lies in contemporary European culture and social disparities. Kumara fitted well with the type of food the bulk of settlers, poor English agricultural labourers, aspired to eat. It was a food that would not have been out of place on the tables of the rich in Britain.

The first opportunity for the English labourer to try such food would have been the 'harvest home,' a tradition in rural England at which landowners provided a large dinner at the end of the harvest with plenty of meat, a food usually absent from most labourers' diets. Another opportunity would arise at weekly provincial markets where local farmers ate a 'farmers' ordinary' at the local inn consisting of '…a thick soup, a pie or savoury pudding, roast meat or poultry, and a sweet pie or pudding with cheese to follow. Unless they had a particularly generous employer, the labourers did not take part… But they were certainly well aware of what they were deprived of…'[50]

The 'farmers' ordinary' of England was almost faithfully transformed into the national meal of New Zealand from early settlement days until the 1970s – the roast dinner. It consisted of roasted meat, roasted root vegetables and tubers, and a boiled green vegetable, flavoured only with salt and pepper and perhaps mint, followed by a pudding.

Another situation in which the less privileged became conscious of dietary differences was on the sailing ships bringing them New Zealand. While on board they would have seen what more prosperous cabin class passengers were served. With little to do but eat, meals served to cabin class emigrants were still rather plain but abundant. Accounts recall enormous meals served in contrast to the tightly regulated monotonous fare doled out to the labourers and their families in steerage.[51]

There are early accounts of missionary wives preparing crystallized kumara and kumara tart in New Zealand but these did not become part of the repertoire of the New Zealand domestic cook.[52] Slow roasted kumara, along with roasted pumpkin, parsnips, onions, and potatoes, all from the settler's own garden, and meat from his farm, meshed perfectly with the colonial settler's idea of good if not luxurious food, and as an ideal has stayed there ever since. It remains an iconic meal with contemporary fast-food outlets specializing in roast dinners that typically include kumara.[53]

I have never heard any New Zealander refer to kumara other than by its Maori name. This suggests that we regard it as integral to our culture. Our national cuisine is no longer represented by a static range of dishes recognizable as the standard fare of a culture 12,000 miles distant. Over the last 30 years New Zealand cuisine has been transformed from the abundant but plain fare of the British agricultural labourer relocated to a South Pacific farm, to a diet that reflects an ever-expanding choice of high quality foods. Most is produced locally, but often prepared with the broadest possible range of international culinary influences. If a national cuisine is defined as an instantly recognizable set of dishes, then New Zealand's is still emerging. With so much recent change in our diet, kumara is a true and enduring product of the *terroir*. It can surely be classed as a national treasure, deserving something like the regard *porcini* in Italy or Puy lentils in France are accorded.

The recipe I have included below reflects these changes. It leans heavily on a food that, thanks to the skills of its Maori guardians, has survived the problems of geographical relocation and remains central to the diet and culture of New Zealanders. The New Zealand Fresh Vegetable Industry website notes that in 2006 New Zealand produced a remarkable 20,000 tons of kumara, marketed fresh and processed in myriad forms. Research into all aspects of commercial kumara production, processing, and marketing continues.[54] Naturally kumara is always part of the New Zealand culinary repertoire for me as a chef helping to promote New Zealand foods abroad. It is also seen on smart restaurant menus throughout New Zealand. Despite a precarious beginning in New Zealand this vegetable has endured and continues to be treasured by generations of both Maori and *Pakeha*. It seems its future as a *taonga* or treasure is 'sweet as,' which in the local vernacular means it will be just fine!

Appendix

SLOW-ROASTED KUMARA SALAD

1.2 kg purple skinned kumara, well scrubbed
4 tablespoons extra virgin olive oil
200g rindless bacon, diced
50 mls cider vinegar
salt and freshly ground black pepper
3 spring onions, thinly sliced
½ cup roasted, unsalted peanuts
¼ cup coriander leaves

Preheat the oven to 200°C
Place the kumara into a large dry roasting dish.
Place in the oven and roast for 1 hour until the kumara is completely soft inside.
Remove kumara from the oven, slice each into quarters.
Place in a large salad bowl.
Heat the oil in a frying pan and fry the bacon until crisp.
Remove from the heat and add the vinegar. Scrape the pan with a wooden spoon and pour over the kumara.
Season the kumara and bacon with a little salt and plenty of pepper.
Sprinkle the spring onions, peanuts and coriander over the top and serve. Serves 4–6.

Notes

1. Walsh 1902.
2. Colenso 1881, p. 9
3. Bassett et al 2004, p.186
4. Furey 2006, p. 9.
5. Bassett et al 2004, p. 186.
6. Furey 2006, p. 23.
7. Bassett et al, p.186.
8. Furey 2006, p. 11.
9. Furey 2006, p. 119.
10. Colenso 1881, p. 11.
11. Bassett et al 2004, p. 186.
12. Leach 2003, p. 452.
13. Furey 2006, p. 12.
14. Colenso 1881, pp. 34–35.
15. Walsh 1902, p. 2.
16. Walsh 1902, p. 3.
17. Lewthwaite 2006, pp. 32.
18. Walsh 1902, p. 4.

19. Furey 2006 p.17
20. Cowan 1930, p. 1.
21. Walsh 1902, p. 7.
22. Colenso 1881, p. 9.
23. Cowan 1930, p. 1.
24. Walsh 1902, p. 5.
25. Moon 2005, p. 62.
26. Walsh 1902, p. 9.
27. Walsh 1902, p. 8.
28. Best 1976, pp. 193–196.
29. Furey, p. 11.
30. Colenso 1881, p. 12.
31. Best 1976, p 171; Walsh 1902.
32. Furey 2006, p. 121.
33. Rubel and Rosman 1971, p. 661; Leach 2003, p. 454.
34. Colenso 1881, p. 18.
35. Rubel and Rosman 1971, p. 662.
36. Roberts 2004, p. 3.
37. Roberts 2004. p. 12.
38. Roberts et al 2004, p. 3.
39. Roberts et al 2004, p. 11.
40. Roberts 2005, p. 5.
41. Roberts et al 2004, p. 8.
42. Roberts 2005.
43. Roberts 2005, p. 1.
44. Roberts 2005, p. 5.
45. Roberts et al 2004 p12.
46. Walsh 1902, p. 14.
47. McGee 2004, p. 305.
48. Walsh 1902, p. 14; Cowan 1930, p. 5.
49. Simpson 1999, p. 90.
50. Simpson 1999, p. 61.
51. Simpson 1999, p. 66.
52. Simpson 1999, p. 90.
53. Lewthwaite 2006, p. 34.
54. Lewthwaite 2006, pp. 34–35.

Bibliography

Bassett, Kari, Hamish Gordon, David W. Nobes, Chris Jacomb. 'Gardening at the Edge: Documenting the Limits of Tropical Polynesian Horticulture in Southern New Zealand.' *Geoarchaeology*, Vol 19, Part 3, pp. 185–218. New York: John Wiley and Sons Inc, 2004.

Best, Elsdon. *Maori Agriculture*. Wellington: A.R. Shearer Government Printer, 1976.

Colenso, William. 1881. 'On the Vegetable Food of the Ancient New Zealanders before Cook's Visit.' *Transactions and Proceedings of the New Zealand Institute*, 1880, Vol XIII pp. 3–39.

Cowan, James. *The Maori Yesterday and To-day*. New Zealand Electronic Text Centre. Chapter XV.

'The Cultivation of the Kumara', 1930. www.nzetc.org/tm/scholarly/tei-CowYest-tl-body-dl5.html Accessed 14 April 2008.

Furey, Louise. *Maori Gardening. An Archaeological Perspective*. Wellington: Department of Conservation, Science and Technical Publishing, 2006. http://www.doc.govt.nz/upload/documents/science-and-technical/sap235.pdf. Accessed 16 May 2008

Leach, Helen. 'Did East Polynesians have a concept of luxury foods?' *World Archaeology.* Vol 34, 2003, pp. 442–457.

Lewthwaite, Steve. 'Sweet Potato Products in a Modern World: The New Zealand Experience.' International Society for Horticultural Science Acta Horticulturae 703. *II International Symposium on Sweetpotato and Cassava: Innovative Technologies for Commercialization*, 2006, pp. 31–37.

McGee, Harold. *McGee on Food and Cooking. An Encyclopedia of Kitchen Science, History and Culture.* London: Hodder and Stoughton, 2004.

Moon, Paul. *A Tohunga's Natural World. Plants, gardening and food.* Auckland: David Ling Publishing Ltd., 2005.

Roberts, Mere. *Walking backwards into the future: Maori views on genetically modified organisms.* World Indigenous Nations Higher Education Consortium, 2005. www.winhec.org/docs/pdfs/Journal/Mere%20Roberts.pdf.

Roberts, Mere, Brad Haami, Richard Benton, Terre Satterfield, Melisssa Finucane, Mark Henare, Manuka Henare. 'Whakapapa as a Maori Mental Construct: Some Implications for the Debate over Genetic Modification of Organisms.' *The Contemporary Pacific*, Vol 16, No. 1, 2004. http://muse.jhu.edu/journals/contemporary_pacific/v016/16.1roberts.html Accessed 5 May 2008.

Rubel, Paula and Abraham Rosman. 'Potlatch and Hakari: An Analysis of Maori Society in Terms of the Potlatch Model.' *Man*, New Series, Vol. 6, No. 4, 1971, pp. 660–673.

Simpson, Tony. *A Distant Feast. The Origins of New Zealand's Cuisine.* Auckland: Random House New Zealand, 1999.

Walsh Archdeacon, Phillip. 'The Cultivation and Treatment of the Kumara by the Primitive Maoris.' *Transactions and Proceeding of the Royal Society of New Zealand 1868–1961.* Volume 35 1902. Art. II. rsnz.natlib.govt.nz/volume/rsnz_35/rsnz_35_00_000510.html Accessed 14 June 2008

The American Pumpkin

Mark McWilliams

In her 1827 novel *Northwood*, Sarah Josepha Hale claimed pumpkin pie was 'an indispensable part of a good and true Yankee Thanksgiving.'[1] As the woman who would become the prime advocate of a national holiday of Thanksgiving, Hale was in a position to define the contents of the quintessential American feast. Yet most authorities agree that there was no pumpkin pie at the first (or second) Thanksgiving; in fact, pumpkin is also notably absent from the Old Colony Club's tradition-setting reenactment in 1769 (which ended with apple pie and 'a course of cranberry tarts').[2] By the mid-1800s, however, pumpkin, whether in pies or on its own, was widely associated with the colonial period, figuring crucially in works from Washington Irving's *The Legend of Sleepy Hollow* to novels like Hale's. In this paper, I trace the use of the pumpkin in the United States and attempt to account for its unlikely inclusion in the iconic American meal.

The pumpkin, like related varieties of squash, originated in the Americas, but was widely disseminated in Europe as part of the Columbian Exchange. Indeed, pumpkins quickly became common in the Old World. Andrew Smith notes that they were 'cultivated in England by the mid-sixteenth century,' and English gardening books from that period consider pumpkins, although they seem less common in cookery books.[3] For example, Keith Stavely and Kathleen Fitzgerald note that 'William Harrison, writing in 1577, found the cultivation of "pompins" […] part of a general resumption, after a few centuries' hiatus, of vegetable growing, not only … among the poor commons, but also among "delicate merchants, gentlemen, and the nobility."'[4]

By the mid-seventeenth century, however, cookery books present recipes for dishes beginning to be recognizable to modern readers as pumpkin pie. In his 1653 masterwork *The French Cook,* La Varenne includes 'Pumpkin Tourte': 'Boil pumpkin in some good milk and strain it through a strainer very thick, then mix it with some sugar, butter, a little salt and, if you like, a few ground almonds' with the mixture baked in a paste.[5] Other cookbooks offer pies of layered sliced pumpkin, often combined with apples or currants. Hannah Wolley's *The Queen-like Closet, or Rich Cabinet Stored with All Manner of Rare Receipts for Preserving, Candying and Cookery* (1670) includes a 'pumpion-pie' of this type: 'Having your Paste ready in your Pan, put in your Pompion pared and cut in thin slices, then fill up your Pie with sharp Apples, and a little Pepper, and a little Salt, then close it, and bake it, then butter it, and serve it hot.'[6] Nath. Brook's *The Compleat Cook, Expertly Prescribing the Most Ready Wayes, Whether Italian, Spanish or French, for Dressing of Flesh and Fish, Ordering of Sauces or Making of Pastry* (1658) gives a more complex preparation:

The American Pumpkin

Take about halfe a pound of Pumpion and slice it, a handfull of Tyme, a little Rosemary, Parsley and sweet Marjoram slipped off the stalks, and chop them smal, then take Cinamon, Nutmeg, Pepper, and six Cloves, and beat them; take ten Eggs and beat them; then mix them, and beat them altogether, and put in as much Sugar as you think fit, then fry them like a froiz; after it is fryed, let it stand till it be cold, then fill your Pye, take sliced Apples thinne round wayes, and lay a row of the Froiz, and a layer of Apples with Currans betwixt the layer while your Pye is fitted, and put in a good deal of sweet butter before you close it; when the Pye is baked, take six yolks of Eggs, some white-wine or Verjuyce, & make a Caudle of this, but not too thick; cut up the Lid and put it in, stir them well together whilst the Eggs and Pumpions be not perceived, and so serve it up.[7]

While cookbooks can be poor guides to the actual fare eaten by their owners, such recipes suggest that pumpkin dishes were not uncommon in England.

The colonists' familiarity with the pumpkin may well account for its relatively quick acceptance into their diet, unlike other native fare that was, at first, rejected. Although pumpkin was a staple for natives in what came to be known as New England, Stavely and Fitzgerald find that 'English observers of native culinary practices had relatively little to say about squashes and pumpkins' (65). Native cooking techniques for pumpkin included roasting and boiling, as well as drying the flesh for preservation. Waverley Root and Richard de Rochemont report natives baking pumpkins whole in the ashes.[8] Even though pumpkin seemed familiar, however, it seems to have taken the threat of starvation to make it palatable. In a period of extraordinary privation, the pumpkin offered sustenance to a desperate people.

Indeed, once adopted, pumpkin seems to have become a staple of the early colonial period. Echoing their provincial defensiveness about the colonial dependence on corn – an attitude that lasted well into the early national period, with both Joel Barlow and Benjamin Franklin eloquently defending the native grain in print – colonists felt moved to defend their reliance on pumpkin as well. In 1654, for example, the New England historian Edward Johnson insisted, 'Let no man make a jest at Pumpkins, for with this fruit the Lord was pleased to feed his people to their good content.'[9] Whether this defensiveness grew from the association of pumpkins and corn with natives the colonists viewed as uncivilized, from the fear of seeming themselves like bumpkins to those back home in England, or, most likely, from a mutually-reinforcing combination of both, the colonists continued to use pumpkins in a wide variety of ways even after other foods became available. In addition to roasted and dried pumpkins, a common preparation involved removing the seeds and filling the cavity with apples, cream, or some combination, effectively producing a kind of self-contained pudding that, in the cream-based version, seems a precursor to the filling for pumpkin pie. Many texts note the use of pumpkin as a sweetener for corn and other breads, and Stavely and Fitzgerald also report pumpkin sauces and pumpkins used to make beer and flip (66–67).

By the end of the eighteenth century, such apologies hardened into defensiveness before melting into both practical advice and nostalgic myth. Practical advice comes in several forms in the first American cookbook, Amelia Simmons's 1796 *American Cookery*. Simmons offers two versions of pumpkin pie that seem clear antecedents of today's Thanksgiving staple:

> No. 1. One quart stewed and strained, 3 pints cream, 9 beaten eggs, sugar, mace, nutmeg and ginger, laid into paste No. 7 or 3, and with a dough spur, cross and chequer it, and baked in dishes three quarters of an hour.
> No. 2. One quart of milk, 1 pint pompkin, 4 eggs, molasses, allspice and ginger in a crust, bake 1 hour.[10]

The second recipe, at least, appears to be for a pie with bottom crust only; this characteristic of the modern pie seems to have become standardized in the opening decades of the 1800s. Simmons also includes a 'Winter Squash Pudding' which she finds 'a good receipt for Pompkins': 'Core, boil and skin a good squash, and bruize it well; take 6 large apples, pared, cored, and stewed tender, mix together; add 6 or 7 spoonsful of dry bread or biscuit, rendered fine as meal, half pint milk or cream, 2 spoons of rose-water, 2 do. wine, 5 or 6 eggs, beaten and strained, nutmeg, salt and sugar to your taste, one spoon flour, beat all smartly together, bake' (46). Published almost exactly contemporaneously with Simmons's cookbook, Joel Barlow's 'Hasty Pudding' (1793) mentions pumpkin as an addition to cornbread, not as a custard, which would seem the logical corollary of hasty pudding, and Royall Tyler's *The Algerine Captive* (1797) mentions pumpkin and milk (as the name of a horse, albeit one evocative of the baked pumpkin custard mentioned in several texts).[11]

Such uses of pumpkins continued to be refined in the first half of the nineteenth century. In her enormously popular *The American Frugal Housewife* (1838), Lydia Maria Child included a lengthy recipe for what seems a quite modern pumpkin pie:

> For common family pumpkin pies, three eggs do very well to a quart of milk. Stew your pumpkin, and strain it through a sieve, or colander. Take out the seeds, and pare the pumpkin or squash, before you stew it; but do not scrape the inside; the part nearest the seed is the sweetest part of the squash. Stir in the stewed pumpkin, till it is as thick as you can stir it round rapidly and easily. If you want to make your pie richer, make it thinner, and add another egg. One egg to a quart of milk makes very decent pies. Sweeten it to your taste, with molasses, or sugar; some pumpkins require more sweetening than others. Two tea-spoonfuls of salt; two great spoonfuls of sifted cinnamon; one great spoonful of ginger. Ginger will answer very well alone for spice, if you use enough of it. The outside of a lemon grated in is nice. The more eggs the better the pie; some put an egg to a gill of milk.[12]

Like Amelia Simmons, Child's contemporaries continued to find a wide variety of uses for pumpkins. In her *Seventy-Five Receipts* (1832), Eliza Leslie noted that stewed pumpkin added to Indian cakes 'is nice' and offered a recipe for pumpkin pudding, which seems like pie filling baked in 'puff paste' shell without a top.[13] Mary Randolph, in her somewhat earlier volume *The Virginia Housewife* (1824), arguably the earliest regional cookbook in the United States, suggested 'Potato Pumpkin,' a baked pumpkin stuffed with forcemeat, and pumpkin pudding, basically modern pumpkin pie filling prepared without a crust.[14] In 1842, Catharine Beecher's *Domestic Receipt Book* included pumpkin bread (a yeast-raised cornbread made with stewed pumpkin), preserved pumpkin, 'New England Squash or Pumpkin Pie,' and 'Mrs. O's Pumpkin Pie' (the real thing).[15]

While American cooks continued to use pumpkins in a variety of dishes, American novelists were developing a more specific use for the pumpkin in their newfound nostalgia for their heritage.[16] At first, writers used pumpkins almost exclusively to evoke the hardships and limitations of the colonial past. Lydia Maria Child's 1824 novel *Hobomok* offers an exemplary passage:

> Breakfast was on the board when I first entered, and after the usual salutations had passed, I with several of my companions, sat down to partake of it. It consisted only of roasted pumpkin, a plentiful supply of clams, and coarse cakes made of pounded maize. But unpalatable as it proved, even to me, it was cheerfully partaken by the noble inmates of that miserable hut.[17]

Although the narrator has just arrived in the New World, surviving the astonishing difficulties of a seventeenth-century Atlantic crossing only to find the 'wretched hovels' of the settlement that would later become Salem, here he rejects his first meal ashore, finding the local fare 'unpalatable.' Specifying the cooking method of the 'roasted pumpkin' emphasizes the unrefined plainness of the meal (as does the careful adjective 'coarse' describing the cornbread). Calling the 'inmates of that miserable hut' 'noble' raises immediate questions for the reader: are they noble because of the mission they've chosen? Because they 'cheerfully' stomach such breakfast? In other words, the narrator suggests that the very attributes that separate the early settlers from the 'civilized' Old World ennoble them. The point here is reinforced much later in the novel, when pumpkin is again used to signal the privations of 'pilgrim fare' (97).

For James Fenimore Cooper, writing at about the same time as Child, pumpkin also stands as symbol, though he uses it to show not the difficulties faced by the early settlers but rather the limitations of life in the imperial provinces. In *The Pioneers* (1823), a long description of dinner fare – unusually specific for fiction of the time – Cooper describes the 'motley-looking pie' made by Remarkable, Judge Temple's cook, 'composed of triangular slices of apple, mince, pumpkin, cranberry, and *custard,* so arranged as to form an entire whole.'[18] Here even pumpkin pie – worlds away from plain roasted pumpkin – marks a lack of refinement. Just like her dessert, Remarkable's dinner seems a particularly American combination of abundance and poor taste; Cooper is careful to show the many unpolished ways in which Remarkable violates the manners and customs of the time. Visiting America in 1842, Charles Dickens perhaps justified Cooper's provincial fears when he found pumpkin part of a similar display of unrefined abundance: 'although there is every appearance of a mighty "spread," there is seldom [much to eat]; except for those who fancy slices of beet-root, shreds of dried beef, complicated entanglements of yellow pickle; maize, Indian corn, apple-sauce, and pumpkin.'[19]

While Dickens's disdain for pumpkin as a side-dish was increasingly common, Cooper's for pumpkin pie stands out of the mainstream. While pie on the frontier may seem rustic when compared to the cuisine of New York society, it still seems about as refined as pumpkin can get. An 1833 story in *The New England Magazine*, for example, declared pumpkin pie one of the 'luxuries suitable for the [Thanksgiving] feast,' and this description seems more representative of the changing reputation of the pumpkin as the nineteenth century progressed.[20] By mid-century, pumpkin seems to disappear from almost all preparations save the increasingly ubiquitous pie.

Indeed, pumpkin pie seems to ascend to iconic status even as most other uses of it fade. While this development may seem strange, several characteristics of the period suggest an explanation. The opening decades of the nineteenth century mark an age in which Americans seem to discover that their young nation has a history, that the colonial past can be usefully reimagined as the crucible of their republic rather than as something to be forgotten. In such a context, even the Starving Time, the period of tremendous suffering faced by the first settlers, can be recast in as positive light as a test passed by the sturdy forebears. In effect, the novelists and writers of the 1820s, '30s, and '40s transform pumpkin – particularly in the refined form of pumpkin pie – from a symbol of difficulty to a link to the good old days. (Here there is a strong analogy to the transformation of the cultural meanings of corn and molasses, although both those foods benefited from explicit political uses – as ways to overcome the British blockade – during the Revolution.[21]) Pumpkin pie, given its literal transformation away from the roasted or dried pumpkin of the Native Americans, was particularly suited for such nostalgia, especially as it became more refined as sugar became more common with increasing economic growth in the United States.[22] Given this history, the best justification for the pumpkin's elevation to its privileged status as one of the essential foods of the national feast seems to be nostalgia.

As a result of this process, Hale was hardly alone in proclaiming pumpkin pie the only fitting end to the Thanksgiving feast. Writers like Cornelius Mathews referred to 'the great Thanksgiving pumpkin' and 'the great Thanksgiving pumpkin pie'; Matthews's nostalgia seems clear in his setting on 'the Old Homestead' located 'in the heart of one of the early states of our dear American Union.'[23] John Greenleaf Whittier's only novel, *Leaves from Margaret Smith's Journal* (1849), may offer an instance of nostalgia's power to transform the past: suppers of roasted pumpkin and milk seem to fit the 1670s, when the novel takes place, but the fondly mentioned pumpkin pies are either anachronistic or refer to something rather different than the contemporary reader would assume.[24] By 1857, Amadis De Gaul found pumpkin pie (along with the newly created mint julep) to be one of the defining features of 'the age we live in.'[25]

(In addition to nostalgia, there may well be unexpected aspect to the pumpkin's adoption as an 'indispensable' part of the national meal. From its first appearances in American literature, the pumpkin has been carried in the strong gothic current in the American mainstream. From its earliest treatment in Irving's *Sleepy Hollow* to modern manifestations in *Scooby Doo* and even the Great Pumpkin, there seems a close association between the globular pumpkin and the supernatural. Certainly it's difficult to discount the close connection between the carved pumpkin's ubiquitous place on Halloween porches and the baked pumpkin pie's ubiquitous place on Thanksgiving tables.)

Such nostalgia secured the pumpkin's place in the American culinary canon, albeit in a limited form that would have surprised the early colonists. Indeed, by the close of the nineteenth century, pumpkin pie had achieved its status as the required conclusion to the Thanksgiving meal, but almost all other uses of the pumpkin had faded away. In her definitive *Boston Cooking-School Cookbook* (1896), Fannie Farmer records this fact without any apparent sense of loss: 'Pumpkins are boiled or steamed same as squash, but require longer cooking. They are principally used for making pies.'[26]

Notes

1. Hale, *Northwood; or, Life North and South, Showing the True Character of Both* (New York: H. Long & Brother, 1852): 87.
2. Old Colony Club. 'A Brief History of the Old Colony Club of Plymouth: Founding of the Club' (2004) 14 April 2008: <http://www.plimoth.org/occhist1.htm>.
3. Smith, *Oxford Companion of American Food* (Oxford, Oxford UP, 2007): 483.
4. Stavely and Fitzgerald, *America's Founding Food: The Story of New England Cooking* (Chapel Hill: U North Carolina P, 2004): 65. Subsequent citations are cited parenthetically in the text.
5. La Varenne, *La Varenne's Cookery: the French Cook ; the French Pastry Chef ; the French Confectioner*, trans. Terence Scully (Totnes, Devon: Prospect, 2006): 317–318.
6. Wolley, *The Queen-like Closet or Rich Cabinet*, 2nd ed. (London: Richard Lowndes, 1672): 208.
7. Brook, *The Compleat Cook* (London: E.B., 1658): 14–15.
8. Root and de Rochemont, *Eating in America: A History* (Hopewell, NJ: Ecco, 1995): 40.
9. Johnson, *Wonder-Working Providence of Sions Saviour in New England* (Delmar, New York: Scholars' Facsimiles, 1974): 56.
10. Simmons, *American Cookery*, reprint (Grand Rapids, MI: William Eerdmans, 1965): 47. Subsequent citations are cited parenthetically.
11. *Harper's New Monthly Magazine* 13:74 (1856): 150.
12. Child, *The Frugal Housewife, Dedicated to Those Who Are Not Ashamed of Economy* (Boston: Carter and Hendee, 1830): 70.

13. Leslie, *Seventy-five Receipts For Pastry, Cakes, And Sweetmeats* (Boston: Munroe and Francis, 1832): 21–22.
14. Randolph, *The Virginia Housewife: or, Methodical Cook* (Baltimore: Plaskitt, Fite, 1838): 108, 127.
15. Beecher, *Miss Beecher's Domestic Receipt Book: Designed As A Supplement To Her Treatise On Domestic Economy* (New York: Harper, 1850).
16. For a much more detailed discussion of this point, see my 'Distant Tables: Food and the Novel in Early America,' *Early American Literature* 38.3 (Winter 2003): 365–393.
17. Child, *Hobomok, Hobomok and Other Writings on Indians,* ed. Carolyn L. Karcher (New Brunswick, NJ: Rutgers UP, 1986): 9. Subsequent citations are cited parenthetically.
18. Cooper, *The Pioneers,* ed. Robert Clark (London: Everyman, 1993): 87.
19. Dickens, *American Notes,* quoted in *Food: An Oxford Anthology*, ed. Brigid Allen (Oxford: Oxford UP, 1994): 228.
20. 'Uncle Sam and His Boys: A Tale for Theologians,' *The New England Magazine* 5:5 (November 1833): 417.
21. See 'Distant Tables' for the details of this claim.
22. In 1854, a magazine article argued that shifting to sugar to sweeten pumpkin pies following the Revolutionary War was a mark of increasing refinement in America ('Art of Eating,' *Putnam's Monthly Magazine* 4:24 (1854): 589).
23. Mathews, *Chanticleer; a Thanksgiving Story of the Peabody Family* (Boston: B.B. Mussey & Co., 1850): 21, 130, 10.
24. Whittier, *Leaves from Margaret Smith's Journal in the Province of Massachusetts Bay 1678–1679* (Boston: Ticknor, Reed and Fields, 1849): 87, 61.
25. De Gaul, 'The Age We Live In, No. 1 (By One Who Doesn't Much Admire It),' *The United States Democratic Review* 40: 2 (1857) : 175.
26. Farmer, *The Boston Cooking-School Cookbook.* Boston: Little Brown, 1896: 268.

Wild Thing: The Naga Morich Story

Michael Michaud and Joy Michaud

Introduction
Early in the twenty-first century an irrevocable change took place in the world of chillies. A series of reports from India, America, and Britain proclaimed that varieties from Bangladesh and north-east India were the most pungent ever measured. Pungency levels of these chillies were more than 50 per cent higher than that of the previous record holder, making this part of Asia a 'haven of heat.' Though the varieties are known by different names and originate in different countries, they are probably part of the same traditional variety that has been growing in this region for years. Naga Morich, the version from Bangladesh, is a popular and indispensable spice among Bangladeshis living in Britain, and appreciation for this chilli is gradually spreading to a wider group of consumers who like their food extremely pungent.

The quest for pungency
Originating in the New World tropics, peppers have become ubiquitous culinary essentials now grown not only in warm climates but in temperate areas as well. They include both sweet and hot varieties, and, botanically, belong to the genus *Capsicum*. Of the more than 20 *Capsicum* species, only five of them have been domesticated (Bosland and Votova).[1]

Renowned for their pungency, chillies receive far more attention and publicity than sweet peppers. The reason for this is difficult to explain, but may have to do with the existence of thousands of varieties whose fruit offer a mind-boggling choice of flavour, shapes, and sizes. Since they adapt well to most of the world's cuisines, it is probably safe to say that there is only a minority of adults in the world who have not eaten chillies at least once in their lives.

The pungency in chillies is caused by a family of chemicals called capcaicinoids. These chemicals are not found anywhere else in the plant world, and the more concentrated they are, the more pungent the chilli is. The concentration of capciacinoids is expressed in its own unit of measurement called the Scoville Heat Unit (SHU), named after Wilbur Scoville, an American pharmacologist who devised a test for measuring chilli pungency in 1912.

Though growing conditions and fruit maturity can significantly affect capcaicinoid concentrations, it is the genetic make-up of the plant that exerts the strongest influence. And since chilli varieties are genetically distinct from each other, they display differences in pungency that range from the barely detectable in some varieties to the virtually mouth-numbing in others. For example, Joe E. Parker, a large Anaheim type, has been measured at a mild 800 SHU and is at the lower end of the range (Bosland et. al). In contrast, the cayenne variety Super Chile (sic) can measure 36,100 SHU (Michaud and Michaud, unpublished data), a level that most people find very pungent. Even more pungent is Rooster Spur, a little-known variety whose tiny fruits reach an astonishing 179,300 SHU (Michaud and Michaud, unpublished data). As pungent as Super Chile and Rooster Spur are, however, they pale in comparison to the habanero variety Red Savina, which once achieved a mouth-numbing 577,000 SHU and for years was the most pungent chilli ever measured.

The hottest chilli in the world has become the Holy Grail among culinary thrill-seekers, and finding it has become a quest bordering on the obsessive. The search seemed to have come to an end with Red Savina, which reigned as the chilli king for so long that it was doubtful if anything would surpass it in pungency. The situation, though, changed in 2001 when scientists working for the Defence Research and Development Establishment in India identified a variety of chilli that measured 855,000 SHU. In a paper reporting their work, the scientists stated that the chilli, which they called Tezpur variety Nagahari, came from the north-east state of Assam and 'seems to be the hottest chilli known so far' (India Academy of Sciences, website).

The Indian discovery, however, was only the first in a series of findings that has led to the recognition of Assam and Bangladesh as havens of extremely pungent chillies. For example, in April 2006 we reported that a chilli we grew measured 923,000 SHU (www.peppersbypost.biz). The chilli, which we named Dorset Naga, was developed in our Dorset market garden from samples of the Bangladeshi chilli Naga Morich bought in Bournemouth. At the same time, we conducted a web search that uncovered Frontal Agritech, an agricultural production and marketing company in the north-east Indian state of Assam. They were selling (and still do) powdered and dried whole fruit of a chilli they call Bih or Naga Jolokia, claiming it reached 1,041,427 SHU (Frontal Agritech, website). And later in 2006, researchers at New Mexico State University in the United States reported that the Assamese Bhut Jolokia they grew measured 1,001,304 SHU (Bosland and Baral).

All of these remarkable chillies are *C. chinense*, and there is enough evidence to suggest that, despite the different names and places of origin, they may be versions of the same traditional variety.[2] In the first instance, they all share extraordinarily high pungency levels. Additionally, the fruit, though not exact replicas of each other, are physically similar enough for the plants producing them to be related. And, finally, parts of Assam and Bangladesh border each other, and their geographic proximity further strengthens the probability that the chillies growing there are, in fact, close relatives.

Plants, especially useful ones, are always on the move, and anecdotal accounts, both verbal and digital, also place this chilli in the Indian states of Nagaland, Meghalaya, and Manipur. They are all close to Bangladesh and Assam, and considering the chilli's wide distribution in the region, it is something of a miracle that its extreme pungency wasn't recognized until the twenty-first century.

The Naga Morich Story

On the move

By the time of the European arrival in the New World, the domesticated species of *Capsicum* were present in the Caribbean, and North and South America. *Capsicum chinense*, the species that includes Naga Morich/Bhut Jolokia (henceforth called Naga Chilli) was no exception and could be found in parts of Central America; on most of the Caribbean Islands; and a large part of South America, including what is now Brazil, Guyana, Surinam, and French Guiana (Andrews).

Before it could became established in Bangladesh and the surrounding Indian states, Naga Chilli first had to make the long journey from its original home to its new residence. When exactly the journey was made and who helped it move are mysteries that will probably never be solved. As A. J. R. Russell-Wood put it: 'Rarely do documentary data permit accurate dating of the first appearance of a specific plant in a region.'

Though Naga Chilli's movements may never be fully understood, it is still of some interest to devise a hypothesis explaining its arrival in Bangladesh and north-east India. Hypothesis-building in this case requires an agent of introduction, and strong contenders for the role must certainly be the early European explorers. Beginning in the sixteenth century, for example, the Portuguese had settlements in Brazil where *Capsicum chinense* was already established (*New Encyclopædia Britannica*, 2003) as well as in or near what is now Bangladesh and Kolkata (Campos; Swartzberg).[3] Their presence in both regions created a bridge over which Naga Chilli could have passed on its move to the east. The Portuguese have already been credited with introducing chillies to Bengal (Campos), and it is certainly possible one of these was *C. chinense*. How exactly it moved is another matter, though it could simply have been by an adventurous sailor carrying seeds on his journeys between the New World and the Asian subcontinent

Other European powers, such as the Dutch and English, also had settlements in both the New World where *C. chinense* originated as well as in and near Naga Chilli's domicile in Bangladesh and north-east India. Any one of these nations could also have been responsible for moving the chilli from the New World to the Old. Attributing Naga Chilli's movements to a single nation, however, simplifies what is probably a convoluted process of plant transference. Naga Chilli, for example, could have first been carried to one or more of the settlements established by Europeans in various parts of Africa. It could then have been cultivated there before going to the Bangladeshi area, possibly making one or more stops in other European settlements located in the Indian subcontinent. Considering the multi-national nature of the settlements, such a movement, carried out over a number of years, could easily have included two countries or more.

Convincing conclusions are difficult to make with only circumstantial evidence, and it may not have been early Europeans who transported Naga Chilli in the first place. Looking to more recent times, another means of introduction could have been indentured labourers moving between India and the New World. From 1838 to 1917, more than half a million Indians went to the Caribbean (Vertovec) and many of them were from Bengal (Look Lai). Their destinations included Jamaica, British Guiana (now Guyana), and Trinidad, and while the majority took up residence there, some returned home, possibly carrying Naga Chilli with them.

Speculation of this nature assumes that the Naga Chilli has remained unchanged and genetically intact since its introduction to Bangladesh and north-east India. It also implies that there exists a New World version ready to take its rightful place along side its Asian brethren. While its existence may at first seem improbable, research done in the Caribbean strongly suggests that Naga Chilli's New World counterpart may have, in fact, been found. Working with *C. chinense* varieties collected in the Caribbean, Umaharan et. al. analysed the pungency levels in the mature fruit of the varieties Seven Pod, Red Scotch Bonnet, and Scorpion. They measured 750,000, 900,000 and greater than 1,000,000 SHU respectively, suggesting that there is a New World version of Naga Chilli.

Of course, none of these Caribbean scorchers may be Naga Chilli at all. In fact, the Naga Chilli that is known today may not even exist in New World and could, instead, be a hybrid of two or more varieties that were themselves introduced to the region. Chillies naturally cross-pollinate, and new varieties are being constantly produced from such crosses. If this is what happened to Naga Chilli, then its parents could have been grown in close proximity to each other, on a small-holding for example, and then cross-pollinated. The offspring of this cross could then have been selected by some astute farmer and propagated further by seed, which then spread throughout this part of the world.

How Naga Chilli came into existence and found its way to the Bangladesh region will probably never be solved. However, the uncertainty surrounding its past gives it a mystique possessed by few other food plants, and speculating on its travels only adds to this mystique.

Naga Morich moves to Britain

Naga Chilli, in its various guises, is currently recognized as the hottest chilli ever measured. Though its world-wide reputation wasn't established until the twenty-first century, local farmers and cooks would have been aware of its exceptional qualities well before then.

In Bangladesh the Naga Chilli is called Naga Morich, *morich* being the Bangladeshi name for chilli. Over a quarter of a million Bangladeshis now live in Britain, and they have resolutely maintained many of their traditional eating habits, including the consumption of Naga Morich. It is difficult for an outsider to understand the elevated status this chilli enjoys within their community, but as Halima Ahmed, a Bangladeshi woman said to us, 'We are obsessed with it.'

The Bangladeshis not only cook Naga Morich in their food, but they also eat it raw with their meals. Offered a whole fruit at the table, they might, for example, break a

The Naga Morich Story

piece of it off with their fingers. The piece is mixed with a small amount of cooked rice and any other concoctions that are on the plate, and the amalgamation is then popped in the mouth and eaten.

Because Naga Morich is such an important spice for British Bangladeshis, fresh fruits are a common feature in the food shops they patronize. The fruits are distinctively wrinkled and wedge-shaped, and they are almost always sold in the green, unripe stage of maturity.[4]

Some are grown in Britain, e.g. in the London area and in Dorset by us, but the majority are imported. Anecdotal accounts describe a supply chain that has fruit flown from Bangladesh to Heathrow Airport, from where they are collected and distributed as far north as Leeds and as far south as Bournemouth.

Once they reach the shops, the fruits are treated in ways that reflect their special status. For example, they are often sold in a small box or basket near the till, segregated from the other vegetables and chillies. Though chillies are usually sold by weight, Naga Morich is priced individually and can be, therefore, quite expensive to buy.

Commercially made pickles of Naga Morich are an acceptable and popular substitute for fresh fruit. They are a regular feature in shops with a Bangladeshi clientele, who have a choice between at least three brands that are imported and two that claim to be made in England.[5] Though they vary slightly from brand to brand, the ingredients common to all the pickles include finely chopped Naga chillies, vinegar, oil, and salt.

Despite its reputation for being pungent, Naga Morich is not necessarily eaten for its pungency alone. Its appeal may have as much to do with its distinctive aroma, which has variously been compared to apricots and bubblegum. The aroma is typical of *C. chinense* varieties and gives a pleasant flavour to any food to which the chilli is added.

Though shops frequented by Bangladeshis are the traditional sources of Naga Morich, other, newer types of outlets selling both the fresh and processed product are gradually developing. Sold as Dorset Naga, there is a steady demand for fresh fruit delivered through the post by our company, Peppers by Post. Likewise, the supermarket Tesco has shown an interest in Dorset Naga, and in November 2007, they tried selling fresh fruit in their Newcastle store. The response was so positive that they decided to stock it in more of their stores in 2008, using one of their growers, located in Bedford, to produce the fruit.

Internet shopping, too, will uncover at least one brand of Naga Morich pickle that the vendor will post to potential customers, albeit at excessively high shipping rates. Dried fruit, both whole and powdered, are also available on the Internet under Naga's various aliases such as *bhut jolokia*. Alternatively, visits to the ubiquitous Indian takeaways and restaurants scattered all over Britain will frequently uncover dishes made with Naga Morich, probably in the form of a pickle. Bangladeshis normally run these establishments for non-Bangladeshi customers, and the owners are keen to cater to their customers' needs for pungent foods.[6] A typical example is the menu of the Alishaan in Dorchester, which offers a dish called Murgh Zalzala, describing it as '… slightly less

The Naga Morich Story

hot than a vindaloo with Mr Naga! [A very hot chilli pickle.] A fierce dish not for the faint-hearted.'

Naga Morich can be successfully grown in Britain, but like other *C. chinense* varieties, they can be something of a challenge even for experienced gardeners. The seeds are not only slow to germinate, but the plants grow slowly as well. In addition, the first flowers that appear fall off the plants without producing fruit, significantly reducing the yields of the earliest harvests. And because they are ill adapted to a temperate climate, they need to be pampered inside a protective structure like a greenhouse or conservatory, where the climate is warm and more to their liking.

The gardening challenges presented by Naga Morich plants, however, have not deterred green-fingered Bangladeshis from attempting to grow them in Britain. Viable seed is readily extracted from shop-bought fruit, germinated in the warmth, and then nurtured into productive plants. Alternatively, young plants, undoubtedly supplied by enterprising Bangladeshi gardeners, are sold around June in shops located in London and elsewhere. Given the right environment and enough time, these can be prolific enough to make the effort of growing them worthwhile.

Notes

1. *Capsicum annuum* is the most common of the domesticated species. It includes sweet bell peppers and the majority of cultivated chillies, including the cayenne chilli.
   C. chinense includes the hottest chillies ever measured such as the habanero, Scotch Bonnet and Naga Morich.
   C. frutescens includes the world-renowned Tabasco and Thai *prik kii nuu*.
   C. baccatum is known as *aji* throughout South America.
   C. pubescens has black seeds and is called *rocoto* or *locoto* in South America.
2. In their original paper, the Indian scientists called their version *C. frutescens*, which is undoubtedly a misclassification.
3. Kolkata – or Calcutta, as it was once known – is in the Indian state of West Bengal, close to the Bangladeshi border.
4. Fruit of Naga Morich change from green to red as they mature. The preference for green fruit may have to do with their reduced pungency compared to when they are red. According to tests done on Dorset Naga in 2007, green fruit measured 655,900 while red fruit reached 878,400 SHU (Michaud and Michaud, unpublished data). Given these values, the pungency of the red ones may be beyond the tolerance level of even the Bangladeshis.
5. The imported brands include Nicobena's 'Naga King; Pran's 'Naga Pickle'; and The Kitchen Magic's 'Bengali Pepper'. Those produced in England are Shahnaz Food Products' 'Mr Naga' and K&S' 'Naga Raga'.
6. Indian restaurants are found both in large cities and small towns, and they are so prevalent that they have become an integral part of the British culinary scene. Calling them Indian restaurants, however, is incorrect, and as Lizzie Collingham said in her 2006 book *Curry*: 'There are about eight thousand Indian restaurants in Britain and the great majority of these are run by Bangladeshis'. Of the Bangladeshis living in Britain, it can be said that so seldom has such a relatively small group of people had such a large influence on the eating habits of a nation.

Bibliography

Andrews, Jean. *The Pepper Trail*. Denton: University of North Texas Press, 1999.

Bosland, Paul W., and Jit B. Baral. "Bhut Jolokia' – The World's Hottest Known Chile Pepper is a Putative Naturally Occurring Interspecific Hybrid.' *Hortscience*. 42: 222–224. 2007.

Bosland, Paul W., Jaime Iglesias, and Max M. Gonzalez. 'NuMex Joe E. Parker Chile.' *Hortscience* 28: 347–348. 1993.

Bosland, P.W., and E.J. Votova. *Peppers: Vegetable and Spice Capsicums*. Wallingford: CABI Publishing, 2000.

Campos, J.J.A. *History of the Portuguese in Bengal*. Calcutta: Butterworth & Co.Ltd. and London: Butterworth and Co., reprinted 1998. New Delhi: Asian Educational Services, 1919.

Collingham, Lizzie. *Curry*. London: Vintage, 2006.

Frontal Agritech. www.frontalagritech.co.in/download.htm (download pdf document 'Brochure on Bih jolokia or Naga jolokia (Capsicum chinense)')

India Academy of Sciences. www.ias.ac.in/currsci/aug102000/scr974.pdf

Look Lai, Walton. *Indentured labor, Caribbean Sugar*. Baltimore: Johns Hopkins University Press, 1993.

Michaud, Michael and Michaud, Joy. www.dorsetnaga.biz

The New Encyclopædia Britannica. Volume 4: French Guiana. Volume 8: Netherlands Antilles. Volume 20: Guyana. Volume 28: Suriname. Volume 29: West Indies. Chicago: Encyclopædia Britannica, Inc. 2003.

Russell-Wood, A.J.R. *The Portuguese empire, 1415 to 1808*. Baltimore: Johns Hopkins University Press, 1998.

Suttons. *Catalogue*. 2009.

Swartzberg, Joseph E. *A historical atlas of South Asia*. http://dsal.uchicago.edu/reference/swartzberg/ 1992.

Umaharan, Pathmanathan, Adams, Herman and Moses, Marissa. 'Caribbean hot pepper germplasm management.' *Proceedings of the Caribbean hot pepper industry workshop,* Port of Spain, 26–28 November 2003. (available at http://www.cardi.org/publications/proceedings/hotpepper/4.5.html). 2004.

Vertovec, Steven. Indian indentured migration to the Caribbean. In: *The Cambridge survey of world migration,* ed. Robin Cohen. Cambridge: Cambridge University Press, 1995.

'Per rape et porri et per spinachi': Examining the Realities of Vegetable Consumption at the Monastery of Santa Trinità in Post-Plague Florence

Salvatore Musumeci

This paper examines the realities of vegetable consumption at the Vallombrosan monastery of Santa Trinità in late fourteenth-century Florence.[1] Santa Trinità's fiscal administrator, Dom Lorenzo di Guidotto Martini, kept a record of daily expenditure for the years 1360–1363 called the *libro di spese* or book of purchases.[2] This document is located in the Florentine archives, and is extraordinary in its detail and the fact that it has survived these hundreds of years – normally a book such as this would have been transcribed into a larger set of documents, and the daily records themselves destroyed. The *libro di spese* is utilized in this paper to determine, in part, the role of vegetables in the monastery's deliveries of produce from its farms, daily shopping lists, and meal menus. One would expect, perhaps, to see many different kinds of vegetables, nuts, and fruits under cultivation on the monastery's farms. As this paper will show, however, this was not the case. Though the monastery owned several farms, which produced a variety of items it used on a daily basis, the sheer amounts produced were not enough to satisfy the monastery's needs. The farms provided self-sufficiency for the monastery in terms of olive oil, wood and grain, but the monks had to purchase all other items from the markets, simply supplementing what they acquired in Florence with produce from their lands.

Dom Lorenzo's daily book of purchases begins in January 1360 (1359 in the Florentine calendar). The accounts end in July three years later as plague returned to the city.[3] The short three-year period was not one that witnessed any major economic or political upheavals and, as such, provides insight into what might be regarded as normative for Florentine citizens, both secular and ecclesiastic.[4] This time period was, however, part of a longer transition between the Black Death of 1348 when the city's population had dropped by almost a third, to approximately 55,000 inhabitants, and the final quarter of the century when discontent brought about the uprising of the lowest groups of workers, known as the Ciompi revolt of 1378.[5]

Although Santa Trinità was not one of the oldest foundations of the order, the Vallombrosans took possession of it in 1092.[6] While it had been founded outside Florence's original city walls, the construction of new walls had absorbed it (like many of the city's other monasteries) into an expanding neighborhood.[7] Dom Simone Bencini acted as abbot of the monastery during the time covered by Dom Lorenzo's account book, leading the relatively small number of 14 fully professed bretheren (at minimum); however, the monks themselves were only one component of this community. The congregation of Santa Trinità also included nine lay brothers and thirteen hired servants (their numbers and positions changed rapidly and often throughout Dom Lorenzo's tenure at the monastery) who assisted in the day-to-day running of the monastery.

Like most traditional monastic institutions, Santa Trinità survived and prospered because of long-standing endowments that provided rental income, produce, or both.[8] This ensured that the monastery was closely involved with the rural economy in ways that would have been familiar to most Florentines.[9] During the period in question, the monastery of Santa Trinità owned and farmed a range of properties. The data for the properties located on the monastery's lands has been collected through the expenses incurred in their maintenance as noted by Dom Lorenzo; the book of purchases gives details on seven farms: Arcetri, Campora, Ema, Legnaia, Monte, Mugnone, and San Donnino.[10] While it is possible that some sites were exempt from *gabelle* or import taxes, most of the produce that came from the monastic lands left traces in the account book, since these taxes were imposed on any imports (even on goods the monks already owned) which came through the city walls.

Alongside cost of importation or purchase, convenience had to be considered. Imports from the monastery's farms to the kitchen, pantry or cellar needed to be well-planned and well-orchestrated. Fresh foodstuffs would deteriorate unless they were used immediately or stored properly, and the monastery found it difficult to cope with a glut of seasonal produce. There were, however, strategies for storage that overcame the difficulties. The fruit, nut, and legume production of each of the farms is as follows:

Fruit, Nuts & Legumes[11]	1360	1361	1362	1363
Arcetri	DF	CP, DF, DG	n/a	n/a
Campora	n/a	n/a	n/a	CP
Ema	B, BB, P, A, DF, DP	G, P, A, N, W, F	GL, CP, C, A, W, DF	n/a
Legnaia	n/a	n/a	n/a	n/a
Monte	F	DF	DF, W	A
Mugnone	n/a	n/a	n/a	n/a
San Donnino	BB	n/a	n/a	n/a

The farm at Ema was one of the main producers of goods that included beans, broad beans, chickpeas and garlic – all items that could be dried and stored.[12] Ema also had an orchard, which provided pears, a variety of apples, figs, walnuts, cherries, plums, grapes, and *nespole* or medlars. All of these items were shipped to the monastery either fresh or dried.[13] The second most important supplier of fruits and nuts was the farm at Monte. The produce from Monte – figs, walnuts, and apples – provided a steady supply of items in the period covered by Dom Lorenzo's text; the amounts transported from the farm were more or less consistent over the years represented in the book of purchases.[14] This may have been in part because Monte, like Ema, also had the ability to dry fruits, especially figs, in their *fornace* or oven.[15] While there are no recorded deliveries of legumes or pulses from Monte to the monastery, an entry dated 10 December 1362 indicates that this was soon to change.[16] On that day six *staia* of broad beans were purchased so that they could be sown at Monte, thus ensuring a future supply of the legume. In addition to Ema and Monte, the farm at Arcetri provided a small amount of chickpeas as well as an equally small portion of both dried grapes and dried figs.[17] The farm at Campora produced little in the way of legumes and no fruit at all.[18]

The table above shows the only fruit, nut, or vegetable goods produced on the farms. Surprisingly, there are no import entries within Dom Lorenzo's *libro di spese* that indicate herbs were grown, or that vegetables such as lettuces, leeks, spinach, or squash were cultivated on these lands. Nor are there any entries that would lead us to believe that an agricultural investment was made in the future cultivation of such products. This indicates that only those food-producing plants and trees, which required little upkeep and were able to survive or flourish regardless of weather conditions or personal attention from the farmhand, were grown. These deliberate choices about cultivation on each farm allowed the farmhand to focus his attention on caring for produce or plants that were not only more costly and bore a greater risk of failure, but that were also a central component of the monastery's dietary regime and symbolic in terms of the liturgy – items such as grapevines, olives and grains.[19]

The number of farms owned by the monastery, some of which – like San Donnino and Legnaia – were made up of more than one plot of land, suggests a false sense of abundance, variety, and self-sufficiency. In reality, provisioning this modestly sized monastery in the center of Florence required much more than its farms alone could produce. The monks of Santa Trinità had to visit the markets and shops of Florence regularly in order to supplement their daily dietary regime, as well as to ensure that they would have enough to eat and drink and to entertain others. While items such as grains, olive oil, and wood rarely ever appeared in the daily market purchases recorded by Dom Lorenzo, vegetables and herbs were essential to the kitchen and cuisine of the monastery and had to be procured on a daily basis. In essence, the farms of the monastery served as a supplement to the markets and shops of Florence and not the other way around.

Food historians are always quick to mention that the household garden was tax-free and that it was widely utilized by those in the Medieval and Renaissance periods for herbs and possibly even small vegetable patches.[20] This does not seem to have been the case for the monastery of Santa Trinità. There is no mention of a garden within the walls of the monastery in Dom Lorenzo's text, nor are there any entries for the purchase of seeds or plants for maintaining such a garden.[21] Because of this, even everyday kitchen items such as parsley or sage appear to have been bought at the market. A list of herbs purchased by the monastery is as follows:

Herbs[22]	1360	1361	1362	1363	Total
Herbs	39	22	33	23	117
Mint	4	0	0	1	5
Parsley	32	41	32	23	128
Sage	7	7	7	5	25

These items, which did not respond well to changes in climate or weather conditions and required a good deal of personal attention from farmhands, were easily secured at the markets of Florence. The main use for herbs, along with spices, would have been as flavoring agents, though also possibly as ingredients for a salad.

Vegetables available in the market and purchased by the monastery on an almost daily basis varied from cauliflower, leeks, and fennel, to turnips, spinach, and squash. Purchases of salad and onions were also common. Their cost varied greatly depending on both season and availability. A list of all the vegetables purchased by the monastery is as follows:

Vegetables[23]	1360	1361	1362	1363	Total
Beets	8	2	8	2	20
Cauliflower	120	124	109	48	401
Fennel	44	52	31	21	148
Endive	1	0	0	0	1
Garlic	4	4	1	0	9
Leeks	36	24	38	17	115
Lettuce	2	1	0	3	6
Mushrooms	1	0	0	0	1
Onions	22	58	23	10	113
Pastinache	3	6	2	1	12
Radishes	1	2	0	0	3
Rocket	1	1	4	4	10
Salad	43	65	12	26	146
Scallions	1	1	0	0	2
Spinach	39	17	35	15	106
Squash	15	28	19	9	71
Turnips	10	17	11	5	43

'Per rape et porri et per spinachi'

The large amounts of vegetable products consumed fall in line with the dietary requirements that determined the liturgical year at Santa Trinità. These included the incorporation of beans, pulses, and legumes into the monks' diet as a way of providing variety during the leaner months of the year. A list of all the beans, pulses and legumes purchased by the monastery is as follows:

Beans, Pulses & Legumes[24]	1360	1361	1362	1363	Total
Beans	8	10	7	1	26
Broad beans	11	16	10	10	47
Chickpeas	2	5	0	2	9
Lupine beans	1	3	2	6	12
Peas	0	4	8	2	14
Robigle	8	7	4	3	22

These items not only figured into the leaner menus of Advent and Lent but were also part of dishes that appeared throughout the entire year. Their cost varied but proved more stable as these products could be preserved through drying.[25]

The final section of this paper seeks to illustrate how the items procured at market were transformed into meals, reconstructing, at least in part, what vegetable items were eaten in the monastery of Santa Trinità. This is no simple task: only ingredients are noted in Dom Lorenzo's text, not finished dishes, but by combining the fiscal evidence with contemporary and later cookbooks, we can gain some insights into the culinary culture of the monastery.

Vegetables, herbs, and legumes were sometimes served raw, as in the case of salad; but usually they were boiled, especially beans, onions and leeks.[26] When thrown into the cauldron, vegetables added color and flavor to boiling meat, poultry, or fish, and once cooked, they could be extracted and served as a side dish. Notations of these sides of boiled vegetables or legumes appear constantly throughout Dom Lorenzo's text. Sometimes vegetables appear to have been fried. For example, on 2 April 1360 we read that someone purchased leeks – which were cooked with beans – and spinach to be fried or sautéed.[27] Two other entries in Dom Lorenzo's text note that broad beans and herbs, as well as squash, were also prepared in this fashion.[28] Visitors were present at the monastery each time vegetables were fried.

Since the monks of Santa Trinità regularly ate large amounts of meat and poultry, their consumption patterns certainly fell outside the ideals and practices of the early church fathers, who had defined fasting as the complete abstinence from meat and animal products.[29] However, Santà Trinita's monks carefully observed abstinence from meat during the seasons of Advent and Lent. The importance of these periods is evident in the liturgical rhythms of life at the monastery. Dom Lorenzo makes a point to record

'Per rape et porri et per spinachi'

the days of fast in his book of purchases. For example, on 18 February 1360 Dom Lorenzo records a purchase of cauliflower, lasagna, pomegranate and leeks for the first day of the Lenten season, while on 27 November 1361 the items for an aspic of tench were purchased as Advent began.[30] But while there was a shift in the types of items bought during Lent and Advent, there was no corresponding decline in the quantity of food that was consumed. The consumption of meat spiked before the Lenten and Advent seasons and then disappeared. The amount of fish purchased and consumed during these times increased to make up the deficit.[31] In addition, the congregation often ate entire meals composed solely of vegetables and fruits.

A folio sheet inserted at the beginning of the text, clearly written in Dom Lorenzo's hand, is headed *Questa e la vita quaresimale Mccclxii* or 'This is the Lenten life, 1362'.[32] Unusually, it is an account of what the monks actually ate during Lent, rather than the more typical fiscal record of items purchased.[33] The meals listed are as follows:

Sunday	Lunch: herb fritters and chickpeas Dinner: Fresh fish, with an accompanying sauce, salad and a serving of either fried spinach or white radishes
Monday	Cauliflower, salted fish and either lasagna or vermicelli
Tuesday	Fresh fish, peas and fried spinach
Wednesday	Wine, salted fish and rice or white radishes
Thursday	Fresh fish, chickpeas and fried spinach
Friday	Salted fish, cauliflower and lasagna or vermicelli
Saturday	Fresh fish, wine and broad beans

It is unclear if this sheet was representative for every week of the Lenten diet at Santa Trinità, but it seems likely. It certainly overlaps with menus that can be collated from ingredients recorded in Dom Lorenzo's text during the Lenten periods of 1360–1363.[34] These menus show that two meals were only taken on Sunday (with one exception) and all meals were composed of easily sourced local items and traditional preparations: fish, wine, vegetables, fritters and some sort of starch.[35]

'Per rape et porri et per spinachi'

The monastic fast observed at Santa Trinità was communal, it stressed moderation rather than deprivation and it is unlikely that anyone left the table hungry. The Advent diet, like the Lenten diet, followed the same pattern. We can reconstruct a series of Advent menus for 1360–1362 based upon collated data from purchases recorded in the book of purchases. As an example, a reconstructed menu from the Advent season of 1360 is as follows:[36]

Sunday	Lunch: Strong herb fritters Dinner: Salad and radishes
Monday	Cauliflower and lasagna
Tuesday	Salad of bitter herbs, wine, pepper, saffron, and dried grapes and apples
Wednesday	Cauliflower, spinach and spices
Thursday	*Sorra*, rice and almonds
Friday	Cauliflower, *sorra*, *tonnina*
Saturday	Turnip, leeks, spices and broad beans

As indicated, these Advent menus, which included aspics, fried fish, pastas, and vegetables, were very similar indeed to those served during the Lenten period. They also included special items such as *composte*, *sorra*, and *tonnina* that were served to visitors.[37]

In conclusion, vegetable consumption had a very specific place and purpose in the monastic experience at Santa Trinità. While certain types of fruits and legumes were produced on the farms, it is clear that the quantity and variety were certainly not sufficient for the daily needs of the congregation. In part, the lack of storage space meant that perishable and seasonal items like vegetables could not be grown in quantity, and it was simply more practical to buy from the markets on an 'as-needed' basis. The menus of the fasting seasons indicate that vegetables played a major role in the liturgical life of the monastery, as would be expected. Dom Lorenzo's careful records provide a preciously rare and unique glimpse into the diet and, by extension, daily life of Santa Trinità's fourteenth-century monks.

'Per rape et porri et per spinachi'

Notes

1. I would like to thank Robin Musumeci for her encouragement and for graciously reading and commenting upon various drafts of this paper. I would also like to thank Evelyn Welch, Allen Grieco and Ken Albala for their time and assistance during the writing process. Thanks are also due to the history department, especially Robert Barclay, the administration and faculty development committee of the University of Sioux Falls. Portions of this paper, in varying forms, appear in my unpublished doctoral dissertation. Translations, unless otherwise noted, are my own.
2. Dom Lorenzo's *Libro di spese* was first transcribed by Roberta Zazzeri in 2003.
3. For a list of years that either witnessed plague or famine, see especially Herlihy (1967), p. 105. See also Brucker (1969), pp. 26–27.
4. Brucker (1962), pp. 148–193. See also Najemy (2006), pp. 124–155, and Duggan (1994), pp. 31–59.
5. Brucker (1962), pp. 15–16, Brucker (1969), p. 55, Najemy (2006), p. 97, Gottfried (1983), pp. 45–47, and Herlihy (1997), pp. 39–58.
6. Botto (1939), p. 2. See also Saalman (1966).
7. Murphy (1997), pp. 74–75.
8. Mate (1984), 341–354. See also Cohn (1992), p. 42, Cohn (1999), 1121–1142, Brucker (1962), pp. 18–19, and Brucker (1969), pp. 190–191.
9. This relationship between city and countryside has its roots in ancient urban and rural history and was an important part of daily life and provisioning strategies throughout the Middle Ages and early modern periods. See especially Sereni (1961), Amouretti (1999), pp. 79–89. See also Caferro (1994), 85–103, de la Roncière (2005), Pinto (1996) and (2005), pp. 153–195.
10. Six of these seven properties were situated a short distance south of Florence. Arcetri was located along the hillside to the southwest of Florence with Campora nestled on another hillside just to the southwest; the farm at Ema was in the vicinity of San Felice a Ema, south of the city and east of Scandicci (not far from the farm at Campora). Again to the southwest of Florence, Legnaia lay within close proximity to Campora. Located in Santa Maria a Pineta, the farm at Monte was not far from the monastery or city center. While the majority of lands owned by the monks of Santa Trinità were close to the monastery, Mugnone was the closest while San Donnino lay at the farthest distance.
11. Legend: DF: dried figs, CP: chickpeas, DG: dried grapes, BB: broad beans, P: pears, A: apples, F: figs, DP: dried plums, G: grapes, N: nespole, W: walnuts, GL: garlic, C: cherries, B: beans.
12. Ema and the other lands produced basically the same products – broad beans and chickpeas. Ema appears to have been the only farm where both items were grown at the same time along with garlic and beans. Campora and Arcetri both produced chickpeas but not broad beans, while Monte and San Donnino produced broad beans but not chickpeas. For the delivery of broad beans to the monastery from Ema see ASF, Con. Sopp. 89:45, 15v. For the delivery of an unclassified bean to the monastery from Ema see ibid., 21r. For the delivery of garlic to the monastery from the farm at Ema see ibid., 63v. For the delivery of chickpeas to the monastery from the farm at Ema see ibid., 68r.
13. The ability to dry fruits and pulses for storage and preservation purposes is mentioned on 3 September 1362, see ibid., 68v. The same kind of drying apparatus is listed as existing and providing the same service at the farm at Monte see ibid., 39v. For deliveries of fruits from the farm at Ema see ibid., 20v–21r, 44v–45r, 46v, 58r, 73r–74v.
14. For deliveries of fruits and nuts from the farm at Monte see ibid., 19v, 47r, 48r, 74r, 79r.
15. See note 13.
16. ASF, Con. Sopp. 89:45, 77r. Dom Lorenzo's only other two entries that denote sowing at a particular farm are related to Monte and not Ema, an indication of the growing importance of this farm and the potential to utilize Monte's land in order to bolster the supplies that it offered the monastery. For the entries that mention sowing at Monte see, ibid., 32r, 50r.
17. For the delivery of chickpeas from the farm at Arcetri to the monastery see ibid., 49v. With regard to the dried figs delivered to the monastery from the farm at Arcetri see ibid., 16v, 49v. For the delivery of dried grapes to the monastery from Arcetri see ibid., 49v.

18. For the sole delivery of chickpeas from the farm at Campora see ibid., 91v.
19. Montanari (1994), pp. 15–21, 26–30, (1999), pp. 165–177.
20. See for example Cortonesi (1999), pp. 270–271, and Montanari (1979), pp. 309–318, 351–365.
21. Of course it is a possibility that there was already a well-established garden at the monastery, so that seeds and seedlings would have been in perpetual supply. These items would not have then been recorded in the book of purchases.
22. The figures in this chart represent the number of times a product was purchased and recorded by Dom Lorenzo, and do not represent quantity, which was hardly ever stated.
23. The figures in this chart represent the number of times a vegetable was purchased and recorded by Dom Lorenzo, and does not represent quantity, which was hardly ever stated. *Pastinache* are a member of the radish family but are white in color and similar in shape to a carrot. See Frosini (1993), p. 115.
24. The figures in this chart represent the number of times a bean, pulse, or legume was purchased and recorded by Dom Lorenzo, and does not represent quantity, which was hardly ever stated. *Robigle* is a legume that is very similar to a pea in shape and appearance. Frosini (1993), pp. 118–119.
25. Giagnacovo (2002), pp. 116, 118, 120–121.
26. See for example the entry on 4 November 1362, ASF, Con. Sopp. 89:45, 75r. For other similar entries see ibid., 5r, 55v, 72v–73r, 74r, 77v, 78v, 79v.
27. Ibid., 6r.
28. On the frying of broad beans see 19 May 1361, ibid., 31v. On the entry that notes the frying of squash see 19 June 1361, ibid., 34r.
29. Westermarck (1907), pp. 391–422, Montanari (1994), pp. 21–25, 78–82, Scully (1995), pp. 62–64, and Dugan (1995), pp. 539–548.
30. For the example from 18 February see ASF, Con. Sopp. 89:45, 3v. For the example from 27 November see ibid., 48v.
31. Bynum (1987).
32. ASF, Con. Sopp. 89:45, first page of text (no pagination).
33. Ibid.
34. For a menu sample for 1360 see 1–7 March, ibid., 4v–5r. For 1361 see 28 February–6 March, ibid., 27v–28r. For a menu sample for 1362 see 6–12 March, ibid., 54r–54v., and for 1363 see 20–26 February, ibid., 80v.
35. The Lenten menu of 1361 has a Monday wherein two meals were taken. For a period recipe for fritters see Faccioli (1966), p. 71.
36. The menu sample for 1360 comes from 29 November–5 December, ASF, Con. Sopp. 89:45, 23v–24r. For a sample menu from 1361 see 5–11 December, ibid., 49r–49v. And for 1362 see 12–18 December, ibid., 77r–77v.
37. *Composte* is similar to vegetable compote, Frosini (1993), pp. 145–146. *Sorra* is salami made from the belly of tuna, ibid., pp. 99–100. *Tonnina* is salami made from the loin of tuna, ibid., pp. 100–101.

Archival sources
Archivio di Stato di Firenze: ASF, Con. Sopp. 89:45.

Bibliography
Amouretti, Marie-Claire. 'Urban and Rural Diets in Greece.' In Flandrin and Montanari, eds. *Food: A Culinary History*, 79–89.
Botto, Carlo. 'Note e documenti sulla chiesa di S. Trinità in Firenze.' *Rivista d'Arte*, 20 (1939): 1–22.
Brucker, Gene. *Renaissance Florence*. New York and London: John Wiley and Sons, Inc., 1969.
———. *Florentine Politics and Society, 1348–1378*. Princeton: Princeton University Press, 1962.
Bynum, Caroline Walker. *Holy Feast and Holy Fast: The Religious Significance of Food to Medieval Women*. Berkeley and Los Angeles: University of California Press, 1987.

Caferro, William. 'City and Countryside in Siena in the Second Half of the Fourteenth Century.' *The Journal of Economic History*, 54 (1994): 85–103.

Cohn, Samuel. 'Piety and Religious Practice in the Rural Dependencies of Renaissance Florence.' *The English Historical Review*, 114 (1999): 1121–1142.

———. *The Cult of Remembrance: Six Renaissance Cities in Central Italy*. Baltimore: The John Hopkins University Press, 1992.

Cortonesi, Alfio. 'Self-Sufficiency and the Market: Rural and Urban Diet in the Middle Ages.' In Flandrin and Montanari, eds. *Food: A Culinary History*, 268–274.

de la Roncière, Charles-Marie. *Firenze e le sue compagne nel trecento: mercanti, produzione, traffici*. Florence: Leo S. Olschki Editore, 2005.

Dugan, Kathleen. 'Fasting for Life: The Place of Fasting in the Christian Tradition.' *Journal of the American Academy of Religion*, 63 (1995): 539–548.

Duggan, Christopher. *A Concise History of Italy*. Cambridge: Cambridge University Press, 1994.

Faccioli, Emilio ed. *Arte della cucina: libri di ricette, testi sopra lo scalco, il trinciante e I vini dal XIV al XIX secolo*. Milan: Edizioni Il Polifilo, 1966.

Flandrin, Jean-Louis and, Massimo Montanari, eds. *Food: A Culinary History*. Trans. Albert Sonnenfeld. New York: Columbia University Press, 1999.

Frosini, Giovanna. *Il cibo dei signori: la mensa dei priori di Firenze nel quinto decennio del sec. XIV*. Florence: L'Accademia della Crusca, 1993.

Giagnacovo, Maria. *Mercanti a tavola: prezzi e consumi alimentari dell'azienda Datini di Pisa, 1383–1390*. Florence: Opus Libri Edizioni, 2002.

Gottfried, Robert. *The Black Death: Natural and Human Disaster in Medieval Europe*. London and New York: The Free Press, 1983.

Herlihy, David. *The Black Death and the Transformation of the West*. Cambridge MA: Harvard UP, 1997.

———. *Medieval and Renaissance Pistoia: The Social History of an Italian Town, 1200–1430*. New Haven and London: Yale University Press, 1967.

Mate, Mavis. 'Agrarian Economy after the Black Death: The Manors of Canterbury Cathedral Priory, 1348–1391.' *The Economic History Review*, 37 (1984): 341–354.

Montanari, Massimo, 'Romans, Barbarians, Christians: The Dawn of European Food Culture.' In Flandrin and Montanari, eds. *Food: A Culinary History*, 165–67.

———. 'Production Structures and Food Systems in the Early Middle Ages.' In Flandrin and Montanari, eds. *Food: A Culinary History*, 168–77.

———. *The Culture of Food*. Trans. Carl Ipsen. Oxford and Cambridge: Blackwell, 1994.

———. *L'alimentazione contadina nell'alto medioevo*. Naples: Liguori Editore, 1979.

Murphy, Kevin. *Piazza Santa Trinità in Florence, 1427–1498*. Unpublished Ph.D. dissertation. Courtauld Institute of Art, University of London, 1997.

Musumeci, Salvatore. *The Culinary Cultures of the Monastery of Santa Trinità in Fourteenth-Century Florence*. Unpublished Ph.D. dissertation. Queen Mary, University of London, 2008.

Najemy, John. *A History of Florence, 1200–1575*. Malden and Oxford: Blackwell Publishing, 2006.

Pinto, Giuliano. *Toscana medievale: paesaggi e realtà sociali*. Florence: Le Lettere, 2005.

———. *Città e spazi economici nell'Italia comunale*. Bologna: CLUEB, 1996.

Rubinstein, Nicolai, ed. *Florentine Studies: Politics and Society in Renaissance Florence*. Evanston: Northwestern University Press, 1968.

Saalman, Howard. *The Church of Santa Trinità in Florence*. New York: College Art Association of America, 1966.

Scully, Terence. *The Art of Cooking in the Middle Ages*. Woodbridge, Suffolk: The Boydell Press, 1995.

Sereni, Emilio. *Storia del paesaggio agrario italiano*. Roma: Editori Laterza, 1961.

Westermarck, Edward. 'The Principles of Fasting.' *Folklore*, 18 (1907): 391–422.

Zazzeri, Robert, ed. *Ci desinò l'abate: ospiti e cucina nel monastero di Santa Trinità, Firenze, 1360–1363*. Florence: Società Editrice Fiorentina, 2003.

The *Maraîchers* – Market Gardeners of the Ile-de-France

Lizabeth Nicol

> The difference of a single day is perceptible. Vegetables can only be tasted in perfection, gathered the same day.
>
> John Pintard (1759–1844)

Long before we Americans realized that three days of cross-country transport and 30 minutes of boiling would drain the life out of even the heartiest of vegetables, the French knew about the importance of getting vegetables fresh from the garden, quickly to the market, and then to the table. French market stalls groan with dazzling displays of seasonal vegetables. Tiny, slim green beans, feathery fennel, purple baby artichokes, delicate green sprigs of wild asparagus in spring and summer, knobby Jerusalem artichokes and crosnes (*Stachys affinis*), green, orange, and yellow pumpkins in autumn and winter, are all just begging to be taken home and cooked. Local residents fill their wicker baskets and wheeled trolley shopping carts. Restaurant chefs comb the markets in search of the freshest vegetables that will be featured in their daily *menu du marché*. Traditionally served as starters such as *carottes râpés*, or *céleri rémoulade* – grated carrot or celery root salads – or as garnishes for meat and fish, vegetables have gained a new stature on the French menu.

But just from where do all these fruits and vegetables come? They might come from the chef's own garden, as is the case of top Parisian restaurant, l'Arpège. 'We have forgotten that for a very long time, chefs had their own vegetable gardens,' says Chef Alain Passard. He has given vegetables a starring role in his Michelin 3-star restaurant and supplies his own kitchen with modern, heritage and unusual varieties from his gardens to the west of Paris and in Normandy. He is a rare exception. Most chefs purchase their vegetables from Rungis, the huge wholesale market of Paris, as do most of the greengrocers. Rungis gets its produce from all over the world. But a substantial amount of these fruits and vegetables come from the commercial market gardens of the Ile-de-France region.

In his preface to *L'Inventaire du Patrimoine Culinaire de la France – Ile-de-France*, Michel Girard writes, 'In this age, when the fleeting is often seen to be more important than the permanent, where everything modern is exalted, the role of heritage and tradition takes on even more significance. It sets out reference points, markers, of where our cultural roots began. *L'art de vivre*, including, of course, gastronomy, forms an essential part of the heritage of Ile-de-France.'[1] In the past few years, a trend has emerged with more and more chefs returning to the practice of buying directly from

The *Maraîchers* – Market Gardeners of the Ile-de-France

producers themselves. Not only chefs but consumers too want to return to this custom. They want to have contact with the grower, to know the origins of the produce. They want to know how it was grown and be assured of its freshness and quality. They feel more secure in buying directly from the *maraîchers* – the market gardeners – whose gardens seem to have always been a part of the landscape of Ile-de-France.

Ile-de-France and agriculture

> The Ile-de-France was the kernel about which, over a period of nearly a thousand years, the other parts of France gradually gathered.[2]

Agriculture has long been an established feature of the countryside of Paris and its surrounding areas. Gaul was one of the great agricultural regions of the Roman Empire reaching as far north as present day Paris. In 52 BC, Caesar conquered Lutetia, the Latin name for the Paris area, and excavations have revealed traces of settlement gardens and vineyards that grew on the banks of the Seine across from the Ile-de-la-Cité.

Charlemagne took a deep and intelligent interest in the agricultural development of the realm. Around 800 AD, he decreed that certain plants were to be grown in the royal holdings. Helen Morganthau Fox in *Gardening with Herbs for Flavor and Fragrance* states:

> After the fall of Rome, gardens and vegetables are not mentioned at all in the West until 795, when Charlemagne issued his *Capitulare de villis vel curtis imperii*, instructions for the administration of towns under his sway. Charlemagne's edict tells his subjects what he expects of them: 'We desire that they have in the garden all the herbs namely, the lily, roses, fenugreek, costmary, sage, rue, southernwood, cucumbers, pole beans, cumin, rosemary, caraway, chick pea, squill, iris, arum, anise, coloquinth, chicory, animi, laserwort, lettuce, black cumin, garden rocket, nasturtium, burdock, pennyroyal, alexander, parsley, celery, lovage, sabine tree, dill, fennel, endive, dittany, black mustard, savory, curly mint, water mint, horse mint, tansy, catnip, feverfew, poppy, beet sugar, marshmallows, high mallows, carrots, parsnips, oraches, amaranths, kohlrabis, cabbages, onions, chives, leeks, radishes, shallots, garlics, madder, artichokes or fulling thistles, big beans, field peas, coriander, chervil, capper spurge, clary.'[3]

The twelfth century saw the creation of monastic gardens for growing both medicinal herbs and vegetables such as Saint-Opportune Abbey to the north or the hermitage founded to the east of Paris near Meaux by Saint Fiacre who became the patron saint of gardeners.

The appellation Ile-de-France was first mentioned in around 1387 in *The Chronicles* of Jean Froissart, replacing the older name Pays de France.[4]

The *Maraîchers* – Market Gardeners of the Ile-de-France

As Xavier de Planhol states in *An Historical Geography of France*, 'the first mark that Paris made on the agricultural landscape was the early development of a large area devoted to the production of fruit and vegetables.'[5] During the sixteenth century, the French capital experienced a period of sustained population growth and demographic transition. Larger and larger market gardens were needed to provide food for the increasing number of city dwellers. Due to the high cost of land in the city center where this type of cultivation had hitherto been confined, the market gardeners were forced to the swamplands and marshy outskirts or the *marais* where they became known as *maraîchers*.

Agriculture continued to flourish in the seventeenth century with significant developments in horticultural techniques such as experiments with angled glass walls and heating flues to improve the efficiency of greenhouses. In the early 1660s Louis XIV commissioned Jean-Baptiste de la Quintinie to create the kitchen garden at Versailles 'to supply the table of his august master with delicacies all the year round.' De la Quintinie continued to perfect methods of extending seasons and increasing yields through the use of *espalier*, cold frames, and straw mats. His techniques were quickly taken up by the *maraîchers* and evolved into the intensive gardening system.

French intensive gardening

C.M. McKay, a fellow of the Royal Horticultural Society, recorded one of the best descriptions of French intensive gardening as practised by the *maraîchers* of Ile-de-France. Being familiar with the French system, he brought a group of gardeners from Evesham, a middle-sized, rural market town in Worcestershire, to Paris in 1905.

There they saw a highly intensive procedure, using hotbeds, frames, bell glasses, and many specially designed tools and equipment; manure, for example, was carried in baskets on men's backs as paths between the rows were too narrow for a wheelbarrow. Labour-intensive it was, most emphatically; yet it had outstanding merit in calling for far less capital investment in glasshouses and heating systems, and the high productivity of small plots was barely credible.[6]

One of the main differences between ordinary cultivation and intensive cultivation is that in the latter, crops are grown year round in a relatively confined space. Another important aspect is the use of implements to create a microclimate, protect the crops from the elements or to pre-heat the soil before planting, i.e., glass bells or *cloches*, cold frames or straw mats. The aforementioned de la Quintinie perfected many of these methods, but a type of *cloche* was alluded to somewhat earlier in Olivier de Serres' great work, *Le Théâtre d'Agriculture*, which appeared in 1600.

Raised beds of well-turned and well-drained soil, up to 18 inches in height, enriched with horse manure, are also a feature of intensive gardening. Very close plant spacing promotes deep root growth and leaves little room for weeds. The beds are mounded, rather than flattened, to create more surface area for planting. Succession planting becomes a fine art with the second crop being in place as the first is pulled from the

The *Maraîchers* – Market Gardeners of the Ile-de-France

ground, or sometimes the two are planted concurrently with the first crop sheltering the second, which then bursts forth after the harvest of the first.

With their penchant for organization and codifying, the *maraîchers* formed their own guild to set down the principles for their method of gardening. As noted by Aquatias in *Intensive Culture of Vegetables on the French System*, they are: (1) uniformity, (2) constant and even growth, (3) working with and assisting of nature, and (4) intercropping.⁷

Leandre Poisson notes in *Solar Gardening: Growing Vegetables Year-round the American Way*,

> The patterns of labor, the area cultivated, and the size of the heat assisting appliances were all standardized. Frames were five feet long and four and one-half feet wide and bell shaped glass jars or *cloches* were 16 inches high and 17–18 inches in diameter. Many lucrative out-of-season crops were produced by using fermenting horse manure to generate heat under the glass covers. Their system of intercropping was so precise they had no fewer than two and most of the time three crops growing in a bed…harvesting as many as twenty crops a year from a single intensive bed.⁸

In the 1880s, the heyday of Ile-de-France market gardens, there were about 1,800 *maraîchers*, each assisted by four to five laborers, all tilling the soil of the average size holding of about two acres. It has been estimated that there was something on the order of six million *cloches* in use at the time. The Ministry of Agriculture reported an average yield of about 12 tons per holding. At its peak, the *maraîchers* were producing more food than the surrounding population could consume and exports of fresh produce went out to England, Portugal, and the Netherlands.

By the end of the eighteenth century a virtually continuous belt of market gardens surrounded the capital. De Planhol states 'this rural suburban life reached its peak in the nineteenth century, after which it declined under the combined effects of rapid transport and refrigeration. But the decline was slow in coming and only really got under way in the twentieth century. In 1892, the Paris region was growing 20 per cent (in value) of all the vegetables produced in France.'⁹

Ile-de-France and agriculture in the twentieth century

The French farming industry decreased significantly in the twentieth century. Figures from INSEE, the French Statistical Agency, show that the active agricultural population declined from around 10 million in 1911 to 2.5 million in 1981, and today stands at about 1.1 million.

As Henri Mendas and Alistair Cole state in *Social Change in Modern France: Towards a Cultural Anthropology of the Fifth Republic*, 'The extraordinary post-war transformation of French agriculture was spread over a mere thirty years. In 1940 the peasant lived, worked and died in the same village as his ancestors had done since time immemorial, and as deemed to be an eternal creation, imbued with an unchanging timeless soul.'[10] These days the 'businessman' farmer is replacing the traditional farmers with small family-owned farms. The trend towards consolidation has brought about a steady decline in the number of farms from just over two million in 1955 to 545,000 at present.

During the last three decades French farming has undergone dramatic modernization resulting in a marked increase in productivity and yields. The Common Agricultural Policy (CAP) has also had an enormous impact as France is the EU's leading agricultural producer and has greatly benefited from the CAP funds.

Modern day Ile-de-France was created in 1961, forming a ring of eight districts or departments (Essonne, Haut-de-Seine, Seine-et-Marne, Seine-Saint-Denis, Val-de-Marne, Val d'Oise, and Yvelines) with Paris at its center. It extends out over a radius of 50 miles and covers about 4,633 square miles. It is the most populated region in the country with over 11 million inhabitants and is economically one of the richest in Europe with a GDP of over €500 billion in 2006.

Despite its urbanization and economic power, Ile-de-France remains mainly rural with 50 per cent of the land area being used for agriculture, forests and parks. The valleys and vast plains of the Paris basin are endowed with a moderate climate, ample rainfall and fertile soil.

Historically, Ile-de-France has been the *ceinture verte* or green belt where fruits, vegetables, flowers, and nursery plants are grown to supply the Paris region's markets, greengrocers and restaurants throughout the year. However, present day production is overwhelmingly dominated by cereal crops (wheat, corn and barley), sugar beets, and oil seeds (colza and sunflower), destined for both local and international markets. According to *Ile-de-France Destination Saveurs* published by The Centre de Valorisation et d'Innovation Agricole et Alimentaire de Paris Ile-de-France (CERVIA) it represents 87 per cent of overall production.

But, just as in the rest of France, the number of agricultural holdings in the Paris basin is declining at a staggering rate. According to the Chambre Interdépartementale d'Agriculture d'Ile-de-France statistics, there were 14,097 farms in 1970 dropping to 4,190 in 2005 or a loss of nearly 10,000. The decline in numbers of *maraîchers* is even more dramatic falling from 619 holdings in 1988 to 162 in 2000, representing a loss of 457 holdings.

Preserving and protecting the *maraîchers*

Just over a century ago, the majority of the French workforce, around 60 per cent, was involved in agriculture; today it has fallen to an absolute minority of only four per cent. This is viewed as a crisis situation by the French and in keeping with the European Commission's idea of multifunctional agriculture; they enacted the *loi d'orientation agricole* in July 1999. According to Legifrance, this law officially recognizes agricultural policy that 'takes into account the economic, environmental, and social functions of farming and contributes to the management of the territory, with a view to sustainable development.'

The *Maraîchers* – Market Gardeners of the Ile-de-France

In the Ile-de-France, the local *Franciliens* have become increasing alarmed at this rapid loss of farming and the vulnerability of their agricultural traditions. Municipalities, urban planners, and representatives of the farming community as well as local and national agricultural associations are taking action. A new type of agriculture is emerging. It will not only ensure food supply and employment but also recreation, education, social integration, and preservation of traditions. The Conseil Economique et Sociale de la Région Ile-de-France, the Chambre Interdépartementale d'Agriculture d'Ile-de-France, and the Centre de Valorisation et d'Innovation Agricole et Alimentaire de Paris Ile-de-France are among the most active in implementing policies and action plans to protect and revitalize the sector taking into account the new role of agriculture. These include initiatives to improve the quality of agricultural education, facilitating and giving financial support for the integration of young farmers into the community, and measures to enhance the image of agriculture in the eyes of local residents.

Initiatives in 2008 include farmers belonging to the association *Ville-Campagne* meeting with over 8,000 local school children to introduce them to the vocation of farming and to answer their questions. Organized by the Interprofession des Fruits et Legumes Frais (INTERFEL), *Fraich'Attitude* week took place in 2008 from 28 May to 8 June. Through information sessions, tastings, and activities, consumers are encouraged to discover and appreciate fresh produce. Over 12,000 school and office cafeterias participated by featuring menus composed of fresh vegetables.

Terres Fertiles was founded to preserve local agriculture through buying or renting land that will be farmed by the *maraîchers* of the Association pour le Maintien d'une Agriculture Paysanne (AMAP). Their first project is the acquisition of 49 acres of land near Saclay.

The 'back-to-work' program of *Les Potagers de Marcoussis* reflects the social aspect of multifunctional agriculture helping the long-term unemployed by giving them training and work in producing organic vegetables that are sold on a subscription basis to the public.

The cultural and recreational aspects of multifunctional agriculture are demonstrated by the diversification of rural activities, which also help to anchor and sustain small holdings. The rapid development of *gîtes*, bed-and-breakfasts, educational and petting farms, pick-it-yourself growers, and the creation and renewal of markets and various agriculturally related festivals contribute to agricultural education and tourism.

On an international level, AMAP has initiated URGENCI, a transnational project with Portugal and Spain that aims to bring small producers, consumers, and activists together to share experience and information.

Ile-de-France is a member of the Peri-Urban Regions Platform Europe (PURPLE) a pan-European organization which acts as a platform for members to share knowledge, promote successful socio-economic transition and, influence European regional and rural policy making.

The policies of the *loi d'orientation agricole* appear to be having positive results. As Mendras and Cole state in *Social Change in Modern France: Towards a Cultural*

The *Maraîchers* – Market Gardeners of the Ile-de-France

Anthropology of the Fifth Republic, 'After years of stagnation and decline, the countryside is becoming a living entity once again, although it has changed radically in character… local festivals are becoming increasing numerous, and markets and fairs are rediscovering a new popularity.'[11]

Distribution systems

Direct-to-consumer sales, be it to individuals, retailers, restaurants, or wholesalers, comprise short distribution circuits that are of vital importance to the *maraîchers* of Ile-de-France. They account for the distribution of around 96 per cent of production: 54 per cent to wholesalers, supermarkets, restaurants, school and company cafeterias, 24 per cent directly to consumers through local markets, on-line sales and subscriptions, and around 4 per cent for export.

The vast majority of direct-to-consumer sales are through local, outdoor markets which number 78 in Paris and 169 in the Ile-de-France region. However, on-line/subscription sales have been steadily increasing the past few years. AMAPs have grown from 11 in 2005 to 30 today and there are now seven *maraîchers* participating in the Tous Primeurs association. Together, they sell their produce from 41 different sales points.

As consumers have become increasingly concerned about food safety and production, the demand for locally grown produce has risen. In 2007, the Chambre Interdépartementale d'Agriculture d'Ile-de-France and 33 local producers created the *Maraîcher Ile-de-France* logo displayed at farmers' markets and in on-farm shops as a guarantee to consumers that they are buying ultra-fresh, locally grown fruits and vegetables.

Ile-de-France *maraîchers* in the twenty-first century

As quoted in *L'Inventaire du Patrimoine Culinaire de la France – Ile-de-France*, Charles Baltet wrote in 1895, 'The profession of *maraîchers* has always had an honorable and honest reputation, of work well done simply reflecting the general respect with which the profession is held. They are among the hardest working and most thrifty, friends of work and peace.'[12]

Two of the current champions of the cause and among the most active of the *maraîchers* of Ile-de-France are Jean-Pierre Bourven and Joël Thiebault.

Jean-Pierre Bourven is an organic *maraîcher* working near Cergy-Pontoise in the Val d'Oise; he has been called the *Robin de Bio* or Robin Hood of organic growers. At 59 years old, he has been farming since he took over the family holdings at the age of 19 after graduating from the Agricultural School of Jouy-en-Josas. He is a passionate and active defender of organic gardening and serves as Vice-President of the Groupement des Agriculteurs Bio d'Ile-de-France (GAB) and Secretary of the l'Union des Producteurs de Fruits et Légumes d'Ile-de-France.

Jean-Pierre, his wife Evelyne and their three full-time employees and seasonal help maintain 20 acres producing a large selection of organic vegetables, both modern-day and heirloom varieties, as well as large holdings of organic cereal crops. A large part

The *Maraîchers* – Market Gardeners of the Ile-de-France

of their production is sold to Rungis but they also sell directly to the consumer from their on-farm shop, at near-by local markets, and through the AMAP. Depending on the season, one of his *paniers surprises* or surprise harvest baskets might include Argenteuil asparagus, Gournay purple radishes, Paris yellow pumpkins, Ditta potatoes, yellow carrots, or *pain de sucre* lettuce. In discussing *Panier Bio* (Organic Basket), his latest subscription service, he says, 'Every week our customers come to one of the sales points to collect their surprise harvest basket. For me, it's a perfectly satisfying win/win situation. So satisfying that already 84 members have signed up for a quarterly or yearly subscription.'

For Parisian restaurant critic Alexander Lobrano, 'Joël Thiebault, the city's most famous truck farmer, who grows some 1,700 different types of organic legumes on his 54-acre farm in Carrières-sur-Seine, just six miles west of the Eiffel Tower,' is the 'Lord of Legumes' and current celebrity *maraîcher*.[13] His family has been farming in the Yvelines for generations. Along with nine permanent employees and three or four seasonal workers, Joël grows and sells his produce four days a week at the Président Wilson and Rue Gros markets in Paris and by subscription through *Tous Primeurs*. According to the season, shoppers or subscribers will be able to purchase Samos spinach, spring Florence fennel, Azur Star kohlrabi, Cherry Belle radishes, butterhead lettuce, green zebra tomatoes, King Edward potatoes, Hokkaido sweet squash, Guernsey parsnips, and Monarch celery root – among many other varieties.

His vegetables appear on the best tables of Paris. Among his many loyal customers are Michelin-starred chefs Pierre Gagnaire (Pierre Gagnaire), Pascal Barbot (L'Astrance), Hélène Darroze (Hélène Darroze), Jean-François Piège (Les Ambassadeurs), Frédéric Anton (Le Pré Catalan), and Michel Troisgros (La Table du Lancaster). They turn his heirloom vegetables into edible art and are delighted when he brings them something new and different. 'I really enjoy researching and discovering new varieties. This year, it's mizuna, a bit of 'dynamite' for salads. It's a mustard leaf with the peppery taste of roquette and the brutality of mustard – very surprising. Also the *capucine tubereuse* (mashua) – absolutely delicious.' He has come out with the book *Vegetables by 40 Great French Chefs*, which presents the origin and history of 36 of his revived or newly discovered varieties accompanied by a recipe developed by one of his favorite chefs.

When writing about Les Halles many, many years ago, Waverley Root, who discoursed so admirably and lovingly about French and Italian foods, stated:

> …in the 1920s most of the vehicles were horse-drawn farm carts, driven by the same men who had raised the produce and picked it that very morning before daylight so that Paris could have fresh food from the market gardens that encircled it – white beans from Noyon, asparagus from Argenteuil, peas from Clamart, string beans from Bagnolet, cauliflowers from Arpajon, carrots from Crecy. The fields where they grew are buried now under the dormitory suburbs of Paris.[14]

The *Maraîchers* – Market Gardeners of the Ile-de-France

Thus Root laments the loss of the *maraîchers*. I contend that men like Jean-Pierre Bourven and Joël Thiebault prove that the spirit lives on – the *maraîchers* are alive and well and are still living and working the lands of Ile-de-France. These passionate, hard-working women and men merit our respect and admiration and deserve our full support.

Notes
1. Lebey, p. 1.
2. Root, p. 60.
3. Morganthau Fox, p. 45.
4. Froissart, p. 475.
5. de Planhol, p. 259.
6. Thirsk, p. 188.
7. Aquatias, p. 8.
8. Poisson p. 12.
9. de Planhol, p. 259.
10. Mendras, p. 21.
11. Ibid., p. 22.
12. Lebey, p. 115.
13. Lobrano, p. 93.
14. L'Amerloque.

Bibliography
Agreste Ile-de-France. Newsletter Number 82. Montreuil-sous-Bois, France: March, 2006.
Agreste Ile-de-France. La stastique agricole. 2007. <http://agreste.agriculture.gouv.fr>.
ANEFA (l'Association Nationale pour l'Emploi et la Formation en Agriculture). Regards sur l'emploi salarié en agriculture. March 2007. <http://www.anefa.org/>.
AMAP (Association pour le Maintien d'Une Agriculture Paysanne). Carte et Liste des Amap en IdF. 2008. <http//amap-idg.org>.
Aquatias, P. *Intensive Culture of Vegetables on the French System.* 1913. Evanston, Illinois: Adams Press, Reprint, 2007.
ARIA (Association Régionale des Industries Agroalimentaires Ile-de-France). L'Agroalimentaire Francilien. <http://www.aria-idf.net>.
Aubry C., L. Kebir, and C. Pasquier. 'The (Re) Conquest of the Local Food Supply Function by Agriculture in the Ile-de-France Region.' Paper presented at the Second International Working Conference for Social Scientists. Arlon, France: May 28, 2008.

CERVIA (Centre de Valorisation et d'Innovation Agricole et Alimentaire de Paris Ile-de-France). Ile-de-France, Destination Saveurs. Paris, France: February, 2008.

CESR (Conseil Economique et Social Regional Ile-de-France). Le devenir des espaces agricoles et naturels en zone périurbaine. <http://www.cesr-ile-de-France.fr>.

Chambre Interdepartemental d'Agriculture d'Ile-de-France. Le Maraichage (Cultures Légumières). <http://www.ile-de-france.chambagri.fr>.

de Planhol, Xavier and Paul Claval. Trans. Janet Lloyd. *Historical Geography of France*. Cambridge, United Kingdom: Cambridge University Press, 1994.

DRIAF (La Direction Regionale et Interdepartementale de L'Agriculture et de la Foret d'Ile de France). L'Agriculture dans la Ceinture Verte. Rungis, France: DRIAF: 2005.

DIRAF-Ile-de-France, Service des stastiques agricole. L'Agriculture en Ile-de-France. <http://draf.ile-de-france.agriculture.gouv.fr>.

French Ministry of Foreign and European Affairs. The French Economy Over the Last Half Century. 11 August 2005. <http://www.diplomatie.gouv.fr>.

FNPL (Federation Nationale des Producteurs des Legumes). Les Legumes dans la Ville, Dossier de Presentation 2007. <http://www.fnpllegumes.org>.

Fox, Helen Morganthau. *Gardening with Herbs for Flavor and Fragrance* Mineola, NY: Dover Publications, 1933 (reprint, 1970).

Froissart, Jean. trans. Geoffrey Brereton. *The Chronicles*. London: Penguin Classics, 1978.

IAURF (Institute for Urban Planning and Development of the Paris Ile-de-France Region). Un portrait par les chiffres et Chiffres-clés. 2008. <www.iau-idf.fr>.

INSEE (L'Institut National de la Statistique et des Etudes Economiques). Production de quelques produits agricoles. 2008. <http://www.insee.fr>.

INTERFEL (Interprofession de la Filière des Fruits et Légumes Frais). Toute l'activité de la filière fruits et légumes, produits et marchés. <htpp://www.interfel.com>.

L'Amerloque. Commitment. July 09, 2007. <http://amerloqueparis.blogspot.com/2007/07/commitment.html>.

Lebey, Claude ed., *L'Inventaire du Patrimoine Culinaire de la France – Ile-de-France*. Paris: Editions Albin Michel. 1993.

Legifrance, Loi n°99-574 du 9 juillet 1999 d'orientation agricole. <http://www.legifrance.gouv.fr>.

Les Producteurs d'Ile de France en Direct. Producteurs en Detail. <www.producteursidf.com>.

Lobrano, Alexander. *Hungry for Paris*. New York: Random House, 2008.

Mendras, Henri and Alistair Cole. *Social Change in Modern France: Towards a Cultural Anthropology of the Fifth Republic*. Cambridge, United Kingdom: Cambridge University Press, 1991.

Ministère de l'Agriculture et de la Peche. Panorama de l'agriculture et des industries agro-alimentaires. 20 June 2008. <htpp://agriculture.gouv.fr>.

Poisson, Leandre and Gretchen Vogel. *Solar Gardening: Growing Vegetables Year-round the American Intensive Way*. White River Junction, VT: Chelsea Green Publishing, 1994.

PURPLE (Peri-Urban Regions Platform Europe). Mission Statement and About Us. <http://www.purple-eu.org>.

Root, Waverley. *The Food of France*. New York: Vintage Books, 1966.

Terres Fertiles. Le Développement d'une Agriculture Durable en Ile de France. <http://www.terresfertiles-idf.org/>.

Thiebault, Joël, Lyndsay and Patrick Mikanowski. *Vegetables by 40 Great French Chefs*. Paris: Editions Flammarion, 2006.

Thirsk, Joan. *Alternative Agriculture: A History From the Black Death to the Present Day*. Oxford: Oxford University Press, 2000.

Tous Primeurs. Achat en Direct de la Ferme. <http://www.tousprimeurs.com>.

URGENCI. Réseau Urbain – Rural : Générer des Engagements Nouveaux entre Citoyens. <http://www.urgenci.net/>.

The Southern California Vegetable Cult

Charles Perry

A hundred years ago, the American capital of cutting-edge diet was Battle Creek, Michigan, where John Harvey Kellogg's Sanitarium treated every illness with a lacto-ovo vegetarian regime rich in grain foods. Battle Creek's enduring legacy is cold cereals such as cornflakes and granola (a sort of crumbled cookie originally prescribed to patients who were trying to gain weight).

Ten years later, though, Los Angeles had supplanted Battle Creek by boldly turning its diet scheme upside down. The new health gurus exalted fruits and vegetables over all else and were actively hostile to grain, some of them damning it as the principal source of disease. The European fruitarian movement had been anti-grain but not necessarily anti-meat; the Southern California dispensation tended to be vegan as well as grain-free. It regarded fruits and vegetables as numinous, charged with vitality as a battery is charged with electricity.

A cynic might see this as no more than a strategy to delegitimize Battle Creek, since a wide variety of fresh fruits and vegetables were available year-round in Los Angeles but not in frigid Michigan. Also, the Californians tended to prescribe nude sunbathing for all complaints; again, advantage Los Angeles. The real explanation is simpler – the feverish hothouse environment of the area's health-seekers, which also explains the easy confidence with which their food gurus dismissed modern medical science.

Invalids had been moving to Los Angeles for its mild climate – and, at the time, proverbially clean air – since the 1880s. They tended to be Battle Creekers, baking heavy brown bread from unrefined Graham flour and consuming Kellogg products such as Granose, Bromose, and Protose, which one writer described as acting and tasting 'as if it were made of sawdust, tapioca, and hair oil.' (By 1900, there were already cultic manifestations as well, such as the Vegetarian Society of Anaheim.[1] Headed by a mysterious Mr. Thales, its members lived largely on coconuts imported from Tahiti, considering them the perfect food. All the rooms in its headquarters were round – as round as coconuts, presumably.)

At the beginning of the twentieth century, a new element became prominent: retirees from the Midwest. The decisive factor came in the Teens with the emergence of the Hollywood film industry, which created a new and highly addictive form of celebrity. While a top stage actor might hope to be seen by millions of people over the course of a career, millions in a week could adore any Hollywood newcomer.

But looks were all, and the camera is pitiless. Unlike the invalids, who had more or less specific medical complaints, the retirees and the actors were moved by a desperate desire to stave off aging. The diet gurus who appeared promised eternal youth and transcendent vitality. Most of them were seeking the same things for themselves and had moved to Los Angeles in the first place because its climate and fresh produce fit their theories.

The gurus blamed illness and aging on a promiscuous variety of causes: mucus, acidity, fermentation, and the all-purpose villain, devitalized food. By the 1920s, Hollywood health food stores stocked 'alkaline' bread, grain-free uncooked pies, 'non-devitalized' salt (derived from vegetables), and even bubbling oxygenated tooth powder. Such was the local interest that from 1899 to 1940 the *Los Angeles Times* featured a regular column called 'The Care of the Body,' which evenhandedly promoted all the gurus' theories, periodically taking time to deride the ideas that germs cause disease and calories lead to weight gain. The *Times-Mirror Press* also published books by several food gurus.

The grand old man was Otto Carqué (1867–1936), who had begun promoting the healthfulness of local produce in 1905 and by the mid-1920s owned a small chain of health food stores. His extensive writings were filled with scientific tables of the nutritional content of foods, but his thinking was quasi-medieval.[2] Before physicians began to understand the metabolism in the nineteenth century, they had conceived of it as a balance of abstract factors such as hot and cold, moist and wet. For Carqué, it was all about acid versus alkaline, and most medical complaints were due to 'acidosis,' a deficiency of alkalinity in the diet. Conveniently for the California citrus industry, all citrus fruits were richly alkaline. Carqué easily lost patience with people who couldn't see that a lemon, properly understood, was alkaline (his definition of alkalinity being the presence of sodium, potassium, or metallic elements such as iron or copper).

Carqué seems chiefly responsible for the local non-devitalized salt craze. Ordinary table salt, he held, even if it was made from sea salt rather than obviously devitalized rock salt, 'consists of crude particles too coarse and large to be taken up by the blood corpuscles.' (One wonders how his followers didn't notice that salt crystals dissolve in water.) To be assimilated by the body, salt had to be derived from plant products. Others took up the idea during the 1920s and warned that 'inorganic' salt caused rheumatism, eczema, and tuberculosis. Health stores eventually stocked nearly a dozen brands of vegetable salt, including Eka-Salt, Nu-Vege-Sal, Vita-Veg, and Sal-Ray, which was derived from the miracle vegetable celery.

Frank McCoy (1888–1940), who moved to Los Angeles in 1926, was the first diet guru to make a specialty of treating film actors at his clinic near the beach in Santa Monica. Unlike most others, he permitted limited quantities of meat (preferably in the form of Salisbury steak or Belgian hare, and as a result these two items appeared on some Los Angeles menus through the 1950s), though the best foods were green vegetables and non-starchy fruit. He prescribed fasting to treat every complaint, even rheumatism and appendicitis.

Apart from fasting, his great focus was the evil of grain. All starchy foods were suspect, partly because (except for root vegetables) they were devitalized, and it was always dangerous to combine devitalized food with fresh fruits or vegetables, even at a single meal; he considered peach pie a major cause of dysentery. The primary danger was that starches would ferment in the digestive tract, causing toxemia and ultimately cancer. (Unaware that starches are broken down into sugars by digestion, he believed that starch in the diet 'pastes up' the system.) The two worst foods were yeast-risen bread and Graham bread (take that, Battle Creek!), because the yeast organisms are not killed in baking, leading without fail to fermentation in the stomach.[1]

The maddest of the health gurus, whose insane books *Rational Fasting* and *The Mucusless Diet* are still readily available,[4] was Arnold Ehret (1866–1922), who moved to Los Angeles in 1914. Mucus, the source of all disease, was synonymous with constipation and caused by foods that feel sticky – above all grain (the 'paste' specter again), but also meat and dairy products. Ehret accepted green vegetables and non-starchy fruits, though his preferred diet was unfermented grape juice. Still, his ideal was not to eat at all but to take longer and longer fasts until you could dispense with food and live on water, fresh air and sunshine.

In the 1960s, the cover of *The Mucusless Diet* was still a photo of Ehret in his trim Van Dyke beard, staring hypnotically into the middle distance. When the publisher realized that most people buying the book were hippies, it opportunistically substituted a sketch of a man with a luxuriant head of hair, a full beard, and the hurt expression of a misunderstood prophet.

The *Los Angeles Times*'s 'Care of the Body' column was originally written by Harry Ellington Brook, a general assignment reporter with an interest in diet – in time he obtained an N.D. (Doctor of Naturopathy) degree. As early as 1910 he denounced the medical hoax of bacterial contagion and went so far as to claim there was no danger in walking barefoot if one were 'clean inwardly and outwardly' ('Should the hookworm invade, it will not long continue in this uncongenial environment').[5] On his death in 1924, he was replaced by Philip M. Lovell, who continued in the same spirit, insisting, for instance, that a cold is merely the body's attempt to 'cast off toxins' and every fruit, vegetable, and legume is nutritionally balanced in itself.

Lovell had a very successful Naturopathy practice. As a result, he could afford to own several homes, and it is chiefly for them that he is remembered today, because he commissioned the brightest young architects in town to design them. Rudolph Schindler created the Lovell Beach House for him in Costa Mesa (1926) and Richard Neutra the Lovell Health House near Griffith Park (1928). Both buildings, early masterpieces of the Southern California school of modernist architecture, systematically reflected Lovell's belief in the importance of fresh air and sunlight; the Beach House had a private deck for nude sunbathing. Widely publicized when built, they helped create the Los Angeles tradition of erasing, or at least blurring, the distinction between indoor and outdoor, an entirely practical thing to attempt in the city's mild climate.

Because of this architectural trend, in the 1920s, Angelenos began speaking of one's back yard as an 'outdoor living room.' And if it could be a living room, why not a

The Southern California Vegetable Cult

dining room? Or a kitchen? Ironically, the strict vegetarian Lovell had spurred the development of California's emphatically carnivorous backyard barbecue.

The city's last great diet guru was Paul C. Bragg (1895–1976, though he claimed to have been born in 1881). He held that through exercise, deep breathing, fasting, and eating non-devitalized vegetables he had turned his body into an 'ageless, tireless, pain-free citadel of glowing, radiant health' and that anybody who followed his regimen should live to be 120 (at times, he even suggested that it should be possible to live forever). Photos taken late in life show a tall, aggressively vigorous man with a bold shock of white hair and a blazing, faintly wolfish smile.

Bragg's first cookbook had chapters titled 'No Meat' (as the ultimate devitalized food, meat was 60 to 65 per cent uric acid), 'Poisonous White Sugar,' 'Dairy Products Are Not Human Food,' and 'Don't Use Condiments' (meaning spices, vinegar, and pickles, all acid-forming).[6] He was fiercely opposed to bread, of course.

Well, all these gurus were against bread. To Carqué and Bragg, it was acid-forming. McCoy feared intestinal fermentation and blood poisoning. To Ehret, bread was mucus on a plate. But all of them, except Carqué, so far as I can find, eventually pulled in their horns to the extent of allowing brown toast, in which whatever sinister power they saw in bread was tamed. (McCoy insisted that the toast had to be very thin so that the last of the yeast would be killed. Bragg eventually sold his own brand of toasted or 'dextrinized' wheat flour in health food stores; contra Ehret, dextrinized flour was preferable precisely because it was gluey. Ehret himself in time backed off on the idea of an endless fast.)

Bragg recanted even more than the rest. In 1947, the mainstream publisher Alfred A. Knopf issued *Paul Bragg's Health Cookbook*, in which he cautiously accepted meat, though still not when preserved or fried. Bragg no longer considered canned vegetables devitalized in 1947; he even gave recipes for health canning. Eventually he concluded that cider vinegar, one of those forbidden condiments in 1930, was a miracle cure, doubtless shocking all the people who had been dutifully substituting non-devitalized lemon juice all those years.

With Bragg's death, the heroic age of this sort of diet appears to have ended. Many of his books are still in print,[7] and many ideas of the Southern California vegetable cult are still floating around, but no leader of comparable stature has appeared.

What all the gurus had in common was a tendency to seize on some grain (as it were) of truth or sense and exaggerate it into a fever dream of impossible health and security. It may be a testimonial to the more sober sense of our times that the vegetable-based diet has recently reappeared in the California Diet Pyramid, which merely recommends that vegetables, rather than starches, be the foundation of the pyramid. If only I could be sure that my fellow Californians won't eventually take this to the edge of madness again.

Notes

1. Los Angeles Times, June 2, 1901, p. C3.
2. His principal books are *Rational Diet: An Advanced Treatise on the Food Question,* Los Angeles, Times-Mirror Press, 1923; *Natural Foods: The Safe Way to Health,* Los Angeles, Carqué Pure Food Co., 1925; and *Vital Facts About Foods: A Guide to Health and Longevity* , Carqué Pure Food Co., 1933.
3. *The Fast Way to Health*; Los Angeles, Times-Mirror Press, 1923.
4. Amazon stocks new copies of *Mucusless Diet Healing System*, Paso Robles, Calif., Benedict Lust Publications, 1976, and *Rational Fasting*, Benedict Lust Publications, 1971.
5. Los Angeles Times, April 24, 1910, p. IM540.
6. *Live Food Cookbook and Menus*, Hollywood, Live Food Products, 1930.
7. For example, *Bragg Healthy Lifestyle – Vital Living to 120!* (formerly *Toxicless Diet*), with Patricia Bragg, Santa Barbara, Calif., Bragg Health Sciences, 2000; *Apple Cider Vinegar: Miracle Health System*, Bragg Health Sciences, 2002; *Build Powerful Nerve Force*, Bragg Health Sciences, 2008.

From the Plate to the Palate:
Visual Delights from the Vegetable Kingdoms of Italy

Gillian Riley

Italians, in both gastronomy and the visual arts, have enjoyed and celebrated the beauty of vegetables. At one time wild and cultivated plants, grains and pulses, fruit and flowers, nuts and seeds, had been seen as food for animals, the poor, and the not so well off, then tastes changed, and some vegetables joined fruit as part of the diet of the rich and fashionable. They had always been part of the Lenten diet: on the days when the faithful abstained from meat and often from dairy products and eggs, they enjoyed vegetables, raw and cooked, as well as fish. This paper looks at images from the Renaissance onwards to show how what foreigners despised as 'food for bruit beasts' came in Italy to grace the tables of the rich and famous, and hang in gilt-framed splendour as images on their walls.

Food historians have found evidence for this in household accounts and archive material, and even the cookery books of the wealthy, with their profusion of rich meat and fish recipes, also have luxurious vegetable dishes and salads. Naturalists, horticulturalists, physicians, intellectuals, artists, and humble cooks all contributed to a climate of opinion that appreciated and rejoiced in the vegetable world. This paper explores some of the visual and material evidence, and the continuing importance of vegetables in Italian life.

The wonderfully accurate images of fruit and vegetables decorating the Loggia di Psiche in the villa now known as the Farnesina, built in Rome for Agostino Chigi, a rich banker and businessman, were the work of Giovanni da Udine, an associate of Raphael. They were painted in 1518, probably inspired by Chigi's interest in natural history, an enthusiasm shared with his friend the Medici Pope Leo X. The story of Psyche, as told by Apuleius, is illustrated in the vaulted ceiling of the outdoor terrace or loggia. Psyche was a mortal nymph with whom Cupid fell in love, and who after many vicissitudes, including the jealousy of Venus and a brush with the Underworld, was permitted to marry him and join the gods on Olympus. The architectural features separating the vaults of the loggia are decorated with a profusion of fruit and vegetables from the kingdom of Ceres, a celebration of fertility and fecundity, in which the joys of love, procreation and abundance were rammed home in a welter of priapic symbolism. Many vegetables lend themselves to this and were presented in arrangements which leave us in no doubt as to their significance. Both patron and artist had no problem with this conceit of reconciling botanical precision and erotic pagan mythology. Vasari, the painter's biographer, enjoyed the phallic symbolism, and his enthusiasm has distracted some historians from the more mundane merits of these representations of fruit and vegetables. The loggia is best considered in relation to the rest of the villa where every room was decorated with a profusion of gods and goddesses behaving badly and mortals growing old disgracefully, but only here, in an area connecting the interior with the gardens and orchards of the outside world, is this display of pagan voluptuousness framed and adorned with the generous gifts of nature, representations that would eventually become a genre in their own right.

Giovanni da Udine painted plants that had arrived only 20 years before from the New World; he would have been able to study these in the gardens and collections of Rome's intellectual elite with their villas in the tranquil rural area across the Tiber. Chigi, with commercial interests all over Europe and the New World, was well placed to procure specimens. The artist could thus work from life, depicting the fruit and flowers in various stages of ripeness and maturity. There is an attention to detail not always visible from below. (Seeing them on the printed page is a more comfortable experience than craning the neck and squinting upwards towards the vaulted ceiling of the loggia.) We have sightings of aubergines, known in the south but less popular elsewhere; asparagus, by then a popular delicacy; artichokes; a wide range of cucumbers, melons, gourds and squashes; grains, including the recent arrival maize; pulses; broad beans; peas; the new beans from the New World; several varieties of *cucurbita* (marrows great and small); musk melon; and the common bean (*Phaseolus vulgaris*), well before their appearance in published herbals. (Columbus had not met chillies and tomatoes, which accounts for their absence from the loggia.) In addition, there were green vegetables like spinach, chard, a range of cabbages, the beautiful flower heads of fennel and elder, the blossoms of orange, lemon, myrtle and roses, and root vegetables from carrot and parsnip to the various turnips and radishes, as well as salad roots like rampions, nuts of all kinds, and a profusion of succulent fruit, symbols of love and fertility destined for the tables of the rich – peaches, apricots, cherries, apples, pears, azaroles.

There had been earlier detailed observations of fruit and vegetables in paintings, like the apples and cucumbers in the religious works of Crivelli, in the mid-fifteenth century, which are there for their symbolism. In 'The Annunciation' from Ascoli Piceno, apples and cucumbers, far from being the recommended diet of a young pregnant urban housewife, represented the purity of Christ and the fecundity of his mother. Mantegna had used the glowing beauty of citrus fruit in many of his paintings of the Virgin and Child; in the 'Madonna della Vittoria' she sits enthroned under a huge bower of oranges, lemons, limes, and citrons, which could be a reference to local innovations in their cultivation, a sort of agribusiness where capital was needed for the construction and maintenance of the terraces and protective shelters to safeguard this sensitive fruit during the northern winters round Lake Garda.

Caravaggio, whose numinous 'Basket of Fruit' is perhaps the most compelling still life of all, was influenced by artists in the north of Italy who were painting kitchen and market scenes full of fruit and vegetables. Vincenzo Campi's 'Fruitseller' shows a winsome young woman surrounded by fruit and vegetables piled in dishes and baskets, overlapping each other like the dishes in a Veronese banquet. While the bunch of grapes she holds and the pile of white peaches glowing on the green apron on her lap probably have symbolic importance, the majority of fruit and vegetables speak for themselves. They show the contents of the kitchen garden, and indicate how they were appreciated and prepared. The asparagus and artichokes in the bottom left hand corner are set apart from the rest, still expensive luxuries; a close look reveals a bunch of herbs lurking under the artichokes, perhaps the *mentuccia* traditionally used in *carciofi alla romana*. In the right hand bottom corner is a blowsy cabbage, which was both peasant food and, later, an amusing salad for the rich. Fresh young peas are shown shelled, and in their pods (there was also a mange tout variety, eaten when young and tender, with salt and vinegar), they are the luxury, kitchen garden variety, not the dried staple that kept men and beasts going throughout the winter months. Broad beans too are shown as fresh and young, enjoyed then as now, raw with a salty cheese. The pale pink roses scattered over them may have a symbolic role (freshness and purity, a link to the young woman, whose bunch of grapes can indicate both virginity and fruitfulness in marriage). Fronds and bulbs of wild fennel are there, and freshly picked mulberries arranged on a white dish, with a faint blush of juice and a scavenging fly. This is a static, silent composition, in spite of fruit pickers in the background, no buying or selling, no market cries. The fruit and vegetables span the seasons, from the asparagus of early spring to the pumpkins of summer and the cabbages of autumn – most of them food for the rich, not for the market people in Campi's 'Fish Stall' who are enjoying a dish of beans with ribald merriment. Here the artist creates a noisy scene, wriggling fresh fish, vulgar market people, a scavenging cat and barking dog. Some commentators see a connection between the alleged consequences of bean eating and the howling incontinent infant.

Beans are the main item in Carraci's 'The Bean Eater,' in which a man is devouring, rather messily, a bowl of black-eyed beans; he grasps a piece of bread in one none-too-clean hand while slurping a spoonful of beans with the other. The bunch of spring onions is the traditional seasoning of the poor, their pungency a welcome substitute for costly pepper or spices. The man scowls with the urgency of hunger, a battered straw hat still on his head, but he wears a clean white shirt. This is not a totally crude meal, the table is laid with a clean white cloth, the red wine from a nice maiolica jug is served in a glass goblet, a roughly cut tart reveals what could be a filling of ricotta and spinach. Perhaps it was a meal eaten in an inn, the guest a famished countryman.

It was the gastronomic prestige they had among the wealthy, after centuries of scorn and neglect, that made humble vegetables like beans a subject for artists, and salads an inspiration to poets and writers. Once they were prized at the table they appeared on canvas, and by the seventeenth century their portraits were hanging on the walls in gilt frames. Attitudes to vegetables had changed since the Middle Ages, when Albertus Magnus codified the hierarchy of edible plants, a theme discussed by Allen J. Grieco in revealing detail. Roots and pulses were low on the list of vegetable matter, low in their relationship to the earth, food only for the base and the low. In the 'Great Chain of Being,' trees and their fruit were top of the heap, being nearest to heaven, then came arborescent shrubs, shrubs, bushes, vines (trailing and creeping), and then herbaceous plants, with roots lurking well below the ground. Grieco has shown how an analysis of household accounts indicates that vegetables and salads were becoming fashionable in the early sixteenth century.

Vegetables never appeared in aristocratic banquet menus or recipe books of the Middle Ages, even though the poor and less well-off were growing and eating them. By the 1460s Platina, a self-made man, was defensive about his enjoyment of vegetables in the company of other impoverished intellectuals, but they were already creeping into the repertoire of his friend Martino who, cooking for the rich, gives a lovely delicate recipe for fresh broad beans, one for mange-tout peas, a rich cabbage dish, and many uses of spinach and fresh herbs.

About this time the Gonzagas in Mantua were enjoying a small bunch of asparagus, sent to the Duchess, Barbara Hohenzollern, a patron of Platina, in 1446 by her steward: 'Forgive me, there are so few, but I'm sending them since they are a novelty.' In fact artichokes had been known to the Romans and when they were served to Caesar at a banquet in Cisalpine Gaul in 55 BC, his astonishment was at the exotic barbarian sauce of melted butter, not at the tender shoots. We see them being grown in a contemporary *Tacuinum Sanitatis*, a field full of them, harvested by straw-hatted peasants. Almost a century later, in 1530, Isabella d'Este was ordering two cartloads of asparagus to be sent to the Palazzo Te for her self-indulgent son. They appeared in Psyche's loggia in 1518, a decorative bundle of spears wrapped in the broad leaves of flags. Being seasonal and a luxury, they figure in many later still lifes, often alongside artichokes, whose wildly swirling foliage and flower heads lend themselves to decorative compositions.

Artichokes, related to the thistle and cardoon, had been around for some time in the south, introduced from the Arab world in the Middle Ages, but forgotten and unknown in the north. Isabella d'Este, greedy as ever, enjoyed them and demanded supplies of seeds and plants from their agent in Genoa. In 1529, Isabella's son Federigo ordered 300 plants for his kitchen garden. Later in the sixteenth century the great naturalist Ulisse Aldrovandi commissioned illustrations of every kind of animal and vegetable, and among them is an unhappy pet monkey squatting on scarlet silk and grasping a fine plump artichoke, both the animal and the vegetable exotic perquisites of the seriously rich.

The tender leaf base and soft flower base with their pervasive slightly bitter flavour was eventually cultivated in a variety of sizes and colours, and by the mid-seventeenth century Giovanna Garzoni included three kinds in one of the miniatures of fruit and vegetables commissioned by the Medici.

The Grand Duke had inherited the family enthusiasm for natural history that had inspired Pope Leo X (son of Lorenzo the Magnificent), and Garzoni's work was an indication of their interests. Other artists were employed to produce still life paintings as well as scientific work. One of these was Jacopo Chimenti, known to his friends as Empoli. 'Empimi! Empimi!' was his cry, 'Fill me up! Fill me up!' This is just what he did with the good things assembled for the composition of his kitchen still life paintings of the 1620s. A contemporary biographer, Filippo Baldinucci, described how, when he had finished a huge canvas full of flesh, fowl, and a variety of cured and preserved meats, Chimenti would gather friends and assistants in his studio to devour the lot together. The main body of his work consisted of religious subjects which did not offer the same scope for gluttony, although Chimenti showed in them his talent for the accurate depiction of everyday objects. This spirit of scientific accuracy was fostered and encouraged by the Medici rulers of Tuscany in the sixteenth and seventeenth centuries, who were assiduous in their encouragement of agriculture, industry, and commerce, together with the arts that served them. Science was the mainspring of all these endeavours, the inspiration of the artists whose works celebrate the fruit and vegetables that the Medici saw as important contributions to the Tuscan economy. They and other rich noblemen put a lot of energy into both estate management and the collection of rare and exotic plants, and the patronage of botanists and the artists who recorded their achievements. Even in his meat-laden orgies Chimenti allowed the presence of walnuts, a huge cauliflower, radishes and citrus fruits as well as a rope of garlic. Another kitchen scene is almost vegetarian in its scope, with grapes, onions, a serpentine gourd, the *trombetto di Albenga*, radishes, lettuce, figs, peaches, nuts, squash, eggplants, pears, and some ham and cheese and various promising jugs and bottles, more than enough for a fine studio snack.

Noblemen and scientists got together and formed academies for the investigation and discussion of natural phenomena, language, and literature, physics and astronomy. Some of the artists they employed to record their findings were also busy on portraits of fruit and flowers for their ducal patrons. The greatest of these academies was the Accademia Galileiana del Cimento, a collection of scientists and scholars under the patronage of the Medici dukes, who presided over the gatherings at which scientific experiments were conducted and faithfully recorded. It managed to keep alive the ideas and ideals of Galileo even after his disgrace and persecution. Conviviality was a less well recorded aspect of their deliberations, but some of the activities of the Accademia Galileiana del Cimento during the hottest months of the year had practical gastronomical applications – there are numerous mentions of experiments in the freezing of lemon juice, orange blossom water, cinnamon cordial, and other 'waters' or liqueurs.

The arrival from Rome in 1662 of a series of miniatures by Giovanna Garzoni was another source of pleasure to the Medici rulers. These delicately stippled works on parchment are a celebration of the beauties of fruit and vegetables that the painter must have known would appeal to her patrons. Almost in movement on the plate on which they sit, itself balanced in a strange equilibrium on an uneven rocky surface, the stalks of cherries seem to wave at us, while a few already rather dry broad beans contrast with the manic fruit. In another painting peas strive to burst from their pods, dry and ready for keeping, unlike the ephemeral pink and white roses that flop, overblown, on top of the heap. Three different varieties of artichoke sit in a blue and white bowl, surmounted by a pink rose, with a few strawberries on a stem on the ground, expensive seasonal luxuries. Lemons and their fragrant blossoms fill a chipped maiolica dish. A bowl of fresh broad beans reminds one of the pleasure of eating them young with a fresh salty cheese, reminding us how vegetables were prized alongside exotic fresh flowers and luscious fruit.

Garzoni's portraits of citrus fruit and their flowers would have appealed to her Medici patrons. Fruit grown from the grafts and cuttings of the ducal gardeners, evidence of the passion of collectors and the skills of horticulturalists, were used in laboratory and boudoir as well as the kitchen. Later Cosimo would commission vaster works; his physician Francesco Redi had drawing lessons from the artist Filizio Pizzichi who told him of an unknown young painter, Bartolomeo Bimbi, who was doing exquisite flower pieces. Redi showed his work to Cosimo, who commissioned a series of wonderful group portraits of fruit, and some obsessive studies of monstrosities, of which the outsized vegetables are the most appealing to us, putting the gigantism of prize marrows at our village fêtes quite in the shade. These massive mug shots were executed life size, with a caption stating the weight, dimensions, and provenance of the monster. The gastronomic and visual virtues of these are somewhat questionable, presenting a challenge to the artist which Bimbi met by painting the ugly things to fill most of the canvas, as if crushed and confined within the frame, with dramatic lighting and props which conveyed a spurious grandeur, like parodies of heroic ducal portraits. The squash from the Grand Ducal Garden at Pisa is depicted against a wild landscape and a stormy sky, with the leaning tower of Pisa dwarfed in the distance. A cut section lights up the sombre scene with its golden flesh. Fast work was needed with this 177-pound monster, already somewhat deliquescent on its arrival after the long journey to Florence. A contemporary account tells how it was led by the artist to his studio, carried by two strong men, to the applause of an excited crowd of onlookers. A giant cauliflower, glowing within its ring of dark green leaves in the last rays of a troubled evening light, weighed in at 18 pounds, dwarfing its companion, a mighty horseradish, a mere 8 pounds. A crinkly-leafed cabbage is growing in a terracotta pot, on a rocky background, its rather boring foliage lit against a lowering purple sky. An ugly giant beetroot, raised by Filippo Strozzi in March of 1712, still encrusted with the earth from which it had been hewn, is crowned with it foliage, but redeemed by a section cut through the same root, showing coloured rings glowing in the bright spring sunshine. Perhaps the most unprepossessing item in the collection is a giant elderly truffle from Castel Leone, shown whole and in section, weighing four and half pounds, with a caption that gives an all too depressing idea of its rugged state: 'Unlike the soft, smooth interior of a normal truffle this has an uneven, earthy, porous appearance, divided by a multitude of layers, red and white and other colours, with reddish worm-like growths, a strange change in substance wrought over many years.' It is a relief to turn to the golden *ovolo* from Castelfranco di Sopra found on 22 June 1695, 23 ounces of a succulent Caesar's or egg mushroom, of the agaricaciae family, a cut section revealing its beautiful interior, framed by dark wrinkled chestnut leaves with an evening sky beyond.

But Bimbi's major achievements were the huge group portraits of every fruit grown in Cosimo's domains, now to be seen at the villa at Poggio a Caiano outside Florence. Each fruit is numbered and named in a cartouche, amazingly fresh and lively but outside the scope of our examination of vegetables in art.

Salads had also became hugely fashionable, cooked or raw, and even visually boring lettuces and roots were now part of still lifes, genre and kitchen scenes. Salads were originally salt meats or capers, olives, anchovies, and other foods preserved in salt presented in various combinations, hence the name *insalata*. Hot or cold vegetable dishes from fresh vegetables came to be served as salads. (June di Schino has given us an exhaustive account of the joys of the salad, celebrated in literature and history, in her paper to this symposium, on Salvatore Massonio, Giacomo Castelvetro and Costanzo Felici.)

Meanwhile Isabella d'Este delights yet again with her pleasing greed. In February 1519 she wrote to her brother in Ferrara enclosing seeds of a special kind of cabbage, bossily telling him how to prepare it: 'You have to cut away the tough stems and boil the leaves for only a little in water, until the cabbage is just tender, then refresh it in water and dress it with oil and vinegar like a salad. *V. Ex. vederà poi se gli piacerà questa stranieza.*' (And then Your Excellency will see how you enjoy this strange novelty.)

This strange novelty was also enjoyed by her nephew Ippolito in Ferrara. His steward's accounts show the purchase of salads on his journey to France in 1536, so they were by then a rich young man's pleasure, and readily obtainable. Isabella had been assiduous in her pursuit of vegetables – blanched chicory for salads, lettuce seeds from Modena, cabbage seeds, cucumbers and a barrel of pickled fennel from Genoa, white chickpeas, and pumpkin seeds. In 1532 she was demanding *semenza di merizane*, the 'mad apples,' *mele insane* or *melanzane*, a real novelty in the north, described by Costanzo Felice 30 years later as just a pretty face, more of a pleasure for the eye than the palate, but already seen in profusion in Psyche's Loggia in Rome in 1518.

It would be a mistake, though, to assume that salads were a novelty; the poor and middling sort of people had always eaten them. This is seen in an analysis of the household accounts of one of the canny Tuscan merchant Francesco Datini's establishments in Pisa in the late fourteenth century, where expenditure on salads is higher than on any other vegetable. Frugal merchants and monks had always eaten vegetables out of necessity, and salads were everyday fare to them long before they became fashionable enough to become literature.

The reclusive painter Pontormo often ate salads, carefully noted in the journal, begun in 1554, in which he recorded work in progress and his states of mind and body. Influenced by Condivi's recently published biography of Michelangelo he aimed at a prudent moderation in eating and drinking.

Visual Delights from the Vegetable Kingdoms of Italy

Lean or Lenten menus were sketched by Michelangelo on the back of a letter in 1518; they included salt fish, pasta, salad, fennel and spinach, no cheese or eggs, and no mention of fruit. But this was food of some sophistication, not peasant food, prepared with care for a master craftsman who could afford to eat well while working on the rock face at Pietra Santa.

Pontormo's rock face was the chapel in the church of San Lorenzo in Florence, where, elderly and infirm, he toiled with grumpy concentration on the writhing mass of bodies, his homage to Michelangelo's Last Judgement. His diet, over which he brooded as the self-absorbed often do, was exemplary: plenty of good Tuscan bread, local wine, rolled omelettes, salads of lettuce, or borage flowers, or *barba di frate*, stewed peas, broad beans, probably raw in mid March, asparagus in April, cooked radicchio, cabbage, borage stems, sometimes stewed mutton, or bits of offal, or salt fish, occasionally cheese. Sundays he usually ate with Bronzino (who wrote a poem about salads) and his extended family when there would be fresh fish, meat or chicken, and cheerful company; this seemed to do Pontormo good, even if he had no appetite for supper. This modest but pleasant and varied diet cannot possibly be seen as an indication of Mannerist angst or austerity.

Mannerist pomp was to be found in the banquet menus published by Bartolomeo Scappi in 1570, and the patrons of these artists, and possibly Michelangelo himself, might well have been guests at some of them. There over 40 fine cooked vegetable dishes in Scappi's recipe sections, but hardly any mention of salads, which have to be searched for in the hundreds of menus he lists. In the course of the banquets the rich hot food from the kitchen alternated with cold food from the *credenza* or sideboard, allowing discriminating guests to make wise choices, to select a salad of asparagus to follow a rich meat dish, or a few grapes before the boned stewed calves head with garlic and walnut sauce. A meal in the middle of May had seasonal delicacies – raw baby artichokes with salt and pepper, cooked ones with vinegar, and fresh young mange tout peas. A lighter supper the same day included salads of lettuce and blue borage flowers, one of asparagus, a *mescolanza*, and a wonderful salad of sliced citron, dressed with sugar and rosewater, then there were raw and cooked baby artichokes, and *scasi*, sugar snap or mange tout young peas. In the autumn there were salads of carrots, of chicory and of endive; raw cardoons first appeared in October, and truffles, raw and cooked, spinach turned in oil with raisins, capers, and *mosto cotto*. As winter approached there were salads of cooked carrots, cooked onions, and alexanders, along with the usual truffles and roast chestnuts, and soon the crisp roots of chicory and rampions. *Cavoli Bolognesi in minestra con pezzi di cascio per dentro* is a rare sighting of the humble cabbage, and there is a rich dish of turnips cooked with sausages and fennel seeds in January.

Salads as such, in a dish or on a plate, are not often found in art; they usually appear in paintings of Christ sharing a simple meal with disciples. A bowl of green leaves does not have the glamour or quite the same symbolic significance as fruit, (except as the bitter herbs of exile). But salad ingredients often occur in still lifes and kitchen scenes, and many ambiguous bundles of roots and shoots were destined to be eaten raw as salads.

The gastronomic slumming of the Gonzagas and Estes was repeated a generation later when the Grand Duke Ferdinand in Florence commissioned Cristoforo Munari in the early eighteenth century to produce a series of still-life paintings with high and low themes, from bacon and eggs in a terracotta dish, to sponge fingers dunked in wine in the exquisite glassware produced in the Medici factories, rivalling Venetian ware. One of these works could refer to a typical Tuscan dish of pancetta and cabbage, where a rosy pink chunk of cured pork leans like the foreshortened skull in Holbein's 'Ambassadors' against a blowsy cabbage. Another cabbage has a central position in a kitchen scene where the artist's affection for homely domestic pots and pans and ingredients might have reminded the artist's patrons of the cuisine of his native city, Reggio Emilia. Mushrooms, game birds, and live dormice nibbling a leaf of the cabbage hint at an earthy recipe. Cardoons, bread, salame and ham are the subjects of one of a pair of paintings executed for Ferdinando in 1703. Sometimes rustic food is juxtaposed with hot-house fruit and oriental ceramics, another example of the Medici taste for enjoying local products alongside expensive luxuries.

From animal food to elitist luxury, vegetables have been a part of the life and art of Italy for centuries, and it is a pleasure and an education to explore the evidence for this.

The Still-life Painter

Alicia Ríos

Translated by Raymond Sokolov

Photographs by Johanna Hecht

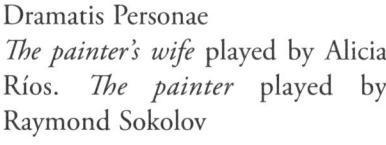

Dramatis Personae
The painter's wife played by Alicia Ríos. *The painter* played by Raymond Sokolov

Setting
Living room – the study of a very poor painter of the 18th Century. There is a table with ready-to-eat vegetable preparations and an easel turned so that the canvas faces away from the audience. To one side is a folding screen.

The wife enters, lugging a basket of vegetables on her right hip. It is as if she had stepped out of Chardin's painting, 'The Return from the Market.' She is bent over from the weight of her basket and moves with effort. Standing straight up with comic relief.

I wonder what kind of still life I can put together from the stuff that happened to be in the market today.

She puts her hand in the basket, staring into space like a magician.

Let's see what rabbits I can pull out of this hat.

The Still-life Painter

She extracts an apple and polishes this shining, tempting fruit. She examines it, sniffs it.

Ah, what an aroma, an edible fruit that looks as opulent as if it had grown on a rose bush!

She places it emphatically on the table. Continuing to probe in the basket like someone who doesn't know what she'll pull out next, she grabs a bunch of leeks.

Leeks of ivory hue, fingers made of onion.

She takes out a cauliflower.

White sponge crowned by a vernal cap.

Next, she takes out a couple of grape clusters.

Tiny bombs of wine juice ready to explode in your mouth.

Finally, and with continuing surprise, she removes a handful of carrots and laughs.

A rabbit's dream.

She turns philosophical.

What conclusion can I draw from this? What unites these things?

She raises the cauliflower over her head with both hands.

To be or not to be? Oh alabaster brains without a skull.

She puts down the cauliflower and picks up two of the leeks, makes a cross out of them and passes it back and forth in the air.

Venerate these bones, these vegetable bones.

She leaves them on the table, still crossed. Then she takes hold of the apple.

A heart that bursts but does not beat.

She brandishes the carrots.

Throats full yellow but not full- throated.

Finally, she caresses the grapes and exclaims.

The Still-life Painter

Tears of Bacchus. He weeps when I squeeze him. *(Almost sobbing)* Arcimboldo, Arcimboldo. Where art thou, Arcimboldo?

Enter the painter looking lost, wearing an 18th-century painter's cap and a greatcoat. He carries a palette of many colors in one hand and a brush in the other.

The wife strikes a sassy pose with arms akimbo and goes toward him with an air of disappointment and scorn.

And I said Arcimboldo?

She looks at him and mugs disrespectfully, like Judy at Punch.

The painter looks at the audience, then at the vegetables, and shrugs.

Okay, so I'm not Arcimboldo. Is that what we're having for dinner?

No, that's what you're going to paint today. And if you were capable of painting faster. If you could manage not to botch them, we could have a stew of leeks and carrots, steamed cauliflower, apple compote, grape syrup so good you'd lick it off your fingers.

I can't. I can't. You've made me so hungry I can't paint.

He approaches the table, licking his chops. She slaps him lightly.

Hold your horses, my little Morandi. As Bertolt Brecht said, Art first, then chow.

He responds by arranging the vegetables in a square, as if he were cooking a dish.

Wouldn't it be better to paint a nice white China plate decorated with the brilliant colors of a well-browned ragout of carrots and leeks? Or an earthenware dish of cauliflower cooked to the point of collapse and giving off curls of steam? Or a crystal beaker with grape jelly glistening inside? *(Pedantically)* These vitreous ceramic textures, their abstract simplicity just don't suit you. Your bag is nature still half alive, still raw, untouched by fire, not half-baked, not ready to eat. *(Changing tone)* Come on, paint, paint. Do something that will make you worthy of this food.

The Still-life Painter

She exits and the painter stays motionless between the palette and the vegetables, in front of the easel, with the brush. At first, he stands between his model and the easel, with the canvas resting on it. Then he starts to examine the 'model' from above, from one side, then from the other.

Form and color. I see them. I can capture them. But what smell, what texture will they have? Will sculpture help?

He abandons his brush and palette and sneaks up on the table like the pink panther and starts to touch and caress the food.

He takes a carrot and passes it under his nose.

What smell will they have? What texture?

He bites off a piece of carrot and begins chewing it like a rabbit.

The texture of a carrot.

He puts the carrot back on the table and picks up the apple.

What aroma will it have? What texture? *He bites into it.* The texture of an apple.

At this point, she re-enters the stage and throws up her hands in despair.

You monster of unbridled gluttony. If I were a painter, I would paint an allegory of appetite out of control.

He picks up an apple and starts to gobble it.

How about this? Don't you think I'm a perfect allegory of appetite out of control?

Enough! We can't go on like this.

She goes behind the screen…and comes back with two dented, rusty cans.

From now on you will paint the tombs of food, sepulchers of the edible tins in the style of Morandi.

The End

But, Did the English Eat Their Vegetables?
A Look at English Kitchen Gardens, and the Vegetable Cookery they Imply (1650–1800)

William Rubel

Of course, the English ate their vegetables. The poor ate them to live – and the rich to do something more than just live; they ate vegetables to do all those many other things people use food for once they no longer have hunger in their bellies. They ate them to pleasure their palates, to show off their social status, to demonstrate culinary hipness,[1] and to be healthy.[2]

Food, medicine, status, and good taste: these were all mixed up in the cuisine of early-modern vegetable cookery. There was not yet a conceptual split between food and medicine. The kitchen garden dispensed plants that both tasted good and had specific health attributes.[3] The spirit with which many high-status diners ate their asparagus is conveyed by the editors of Tournefort's *Great Herbal* (1719):

> Whatever the French aſparagus might be, the English aſparagus, which is the largeſt, fineſt and beſt in the World, both nouriſhes the Body very much, and exhilarates the Spirits very kindly and agreeably.[4]

This is the kind of enthusiastic writing about an ingredient that we now expect to see in cookbooks. But the early-modern cookbook literature rarely includes narrative digressions. This excerpt from Tournefort illustrates how a text from outside the cookbook literature, in this case from an herbal, can help us understand cultural assumptions that are rarely made explicit in cookbooks from this period. The Tournefort text helps us understand not just what vegetables the English ate, but on a deeper level, what they thought about what they were eating.

Cookbooks! Unfortunately, cookbooks are often false friends. Just think of your own cookbook collection. Try to correlate what you actually cook at home and eat against the recipes in your cookbooks. As an experiment, try to deduce from your cookbook collection that a salad of lettuce greens dressed with an oil and vinegar is a possibility in your diet, and if that idea is supported, that you eat it so often it must be seen as an integral part of many meals.[5]

The early-modern cookbook literature is filled with vegetables. Hannah Glasse's important eighteenth-century cookbook, *The Art of Cooking Made Plain & Easy*, mentions approximately 75 vegetables, but these are almost entirely used as ingredients in other dishes, not as foods to be eaten on their own, like a serving of broccoli or plate of micro-greens dressed with oil and vinegar. Seventy-five herbs and vegetables is about the same number of vegetable varieties found in a good California market in June.[6] Even so, this 75 is many fewer vegetables than are mentioned in gardening books from the period, fewer even than the number Hannah Glasse mentions in the kitchen garden section of her own book. Whole classes of vegetables, notably the salad greens, like purslane, which are so much in evidence in gardening books, and in Hannah Glasse's own kitchen garden section, are almost invariably missing from cookbooks. With the notable exception of John Evelyn's book on salads, *Acetaria* (1699), you'd never guess that the affluent English ate green salads, but from the gardening books of the period you'd find it hard to believe they ever sat down to a meal that didn't include one. Substantial portions of the kitchen garden were managed for the production of salad greens, which were understood broadly to include much more than just lettuce.[7]

The absence of salad is just one example of how the cookbook literature alone cannot answer the question of what vegetables were eaten, nor how people felt about them. Salad is the most obvious missing piece to English vegetable cookery, as one can interpret it through cookbooks, but there are other missing pieces, as well. What were done with the pompions – squashes that could include what we now call 'pumpkin'? Pompions were clearly lovingly grown, but how were they eaten? The best culinary information is in herbals, such as Parkinson's *Paradisi in Sole* (1729).

Asparagus is clearly a high-status vegetable. Asparagus recipes are often included in cookbooks. While the most common preparation was probably just boiled asparagus served with butter, Martha Bradley's 'plain asparagus,' Stephen Switzer, in his work *A Compendious Method for Raising Italian Broccoli, Cardoon, Celeriac, and other Foreign Kitchen Vegetables*, explains that the prize part of the broccoli plant are the stalks of the side-shoots which are eaten in lieu of asparagus when it is out of season.[8] While the many gardening calendars, including those in cookbooks[9] include broccoli, cookbooks don't explain their use. From the gardening books it is clear that the production of broccoli was important. Through tricks of the trade, especially succession planting and the use of fermenting dung to keep plants warm during cold weather, broccoli cultivation more or less bridged the natural asparagus season and overlapped with the season for forced asparagus, which by the mid-eighteenth century began as early as November. Producing broccoli side-shoots to substitute for asparagus required space, planning, and skill. It was an expensive enterprise that itself testifies to the status of the broccoli substitute.

But early-modern kitchen gardens, though places for the most intensive agriculture, were more than just agricultural spaces. Before we dig the beds and walk the paths of a kitchen garden, or sit by a fire on a rainy fall evening roasting the sweet root of a skirret in the ashes,[10] or look into the eyes of our friend as he is served the first peas of the season – *ours,* and a full week ahead of *his* – or consider any other aspects of the pleasures of eating vegetables from the early-modern kitchen garden, it will be helpful to situate that garden in the cosmology of the early-modern world. The kitchen garden was conceived as more than just the place for growing vegetables. We live in a more prosaic world. I think it important to keep in mind that there is an otherness associated with the early-modern kitchen garden and its produce. I think that at least some of the time people thought very different thoughts than we do while eating their vegetables, and were thus nourished by them differently. There were two mythical gardens in early-modern England, one 'Eastward of Eden,' and the other in Paradise. Eden's temporal garden was the Kitchen Garden while that of Paradise was the Pleasure Garden.

The wall that surrounded kitchen gardens on great estates, and that at least implicitly surrounded the most humble, marked the physical limits of the garden. So much is easily understood by the modern reader of early-modern gardening books. But what I think is not so easily understood is that the border of the garden also marked the boundaries of a poetic space. This was a space that stood for the time of Innocence when the Garden provided everything Adam needed – food, shelter, and medicine. It is always hard to know the extent to which an idea that is commonly referenced is truly felt, or merely a cliché, but I think that for us, coming from such a different time, that it is helpful to be aware that while the English ate vegetables, as we do, there were ways in which their relationship with their garden and its vegetables might have been different from our relationship to vegetables, and not merely because they often grew theirs, and we don't.

The techniques of the kitchen garden

> Upon the Intervalls 'twixt the lines [of just planted melon seeds] ... you may sow Lettuce-seeds for early sallets, in other Chervill; And you may fringe the whole bed about with purslaine; for these herbs will be very forward, and are to taken up very young...
>
> Bonnefons, *The French Gardiner* (1658)[11]

> But let it be observed, if it is required to have broccoli produce heads before Christmas, that is in October, November, and December; you must sow some seeds of each kind in March, or beginning or middle of April.
>
> Abercrombie, *Every Man his Own Gardener* (1791)[12]

Based on a reading of the cookbook, gardening, and botanic literature, the early-modern kitchen garden was remarkably fecund. Gardening books and gardening calendars often included over 100 different herbs and vegetables, and it seems as if they all were grown. It is my thesis that the fecundity of the kitchen garden drove a vegetable cuisine of unparalleled quality and variety. Malcom Thick, in *The Neat House Gardens*, his work on the market gardens of London, concluded that there was a close correlation between the garden literature and practice.[13] As someone who has used these gardening manuals, in particular Bonnefon's *The French Gardiner* and Abercrombie's *Every Man His Own Gardener*, I am sure that the techniques described work and that the garden calendars can be trusted as statements of actual practice.[14]

But, Did the English Eat Their Vegetables?

You can choose almost any crop – peas, beans, asparagus, purslane, chervil, cauliflower, lettuce, it doesn't matter – and find that the season when the vegetable was available to country house kitchens is vastly longer than one would think possible if one were to judge by the seasons the same vegetables are produced by market gardeners today. Vegetables were also harvested at multiple points in the life-cycle of many plants, suggesting a more finely textured vegetable cuisine, and a more refined level of connoisseurship. For example, spinach was harvested as little plants, in the form of selected leaves of larger plants, and as whole mature plants. Savoy cabbage was harvested as heads of cabbage, but some were permitted to bolt (go to seed), and were then eaten in the form of sprouts which were the side-shoots off the seed stalk.

On the face of it, the works of Bonnefons (1658), Evelyn (1699), Switzer (1731), Langley (1728), Abercrombie (1787), and others describe a gardening system that is so intensive, so costly in terms of material – glass, manure, seeds, and labor – that the garden itself was a sign of the high social status held by culinary vegetables in the second half of the early-modern period (1650–1800). It was the same gardening system practiced by the market gardeners of London and Paris. We can probably assume through an exchange in labor that there was a constant exchange of best practice between the market gardeners and the gardeners on elite estates, not all of which was written down.

We *work* a farm, while we *tend* a garden. Farms are prepared for planting by a superficial scaring of its surface with plow and harrow to depths measured in inches. Gardens are prepared by digging deeply with a spade. In the gardening system described by Nicolas Bonnefons in *The French Gardiner* (1658), beds were dug to depths measured in feet. I emphasize the conceptual difference between farm and garden because that distinction is no longer so clear to kitchen gardeners. For example, the common use today of a rototiller to work the top few inches of garden soil is an example of modeling garden practice after that of the farmer working field crops. Early-modern kitchen gardeners were clear that their gardens were not farms and the literature of kitchen gardening was distinct from that on husbandry. Only for exceptional reasons – for example, the growing of carrots for a commercial seed crop – does one find kitchen vegetables in works on farming.[15]

The early-modern kitchen garden was planted in raised beds that were approximately four feet wide and many yards long. There was usually a pattern to the way in which the beds were laid out. There could be a fountain or a sculptural element to mark a central axis. The use of raised beds was not new to the kitchen gardens of the second half of the early-modern period.[16] What may have been new, and was certainly new in book form, was the use of vast quantities of aged horse manure to create beds of extraordinary fertility, and the use of fermenting beds of manure in conjunction with glass frames to create 'hot beds' to extend the growing season.

For those who could afford it, through deep digging,[17] copious amounts of mulch and manure dug into the soil, and copious amounts of aged horse manure applied as a top dressing,[18] along with an integrated system of more specialized soil amendments, including ditch leaves[19] and manure teas, gardeners created soils of unparalleled (and unnatural) fertility. Whenever I have got close to the amounts of manure specified in *The French Gardiner*, the garden beds have effectively become worm beds.

Garden beds were never walked upon after being formed. Permanent paths ran along the sides of the beds, which were never so wide one could not reach to the middle while kneeling to one side. Deep beds of fertile, uncompacted soil are the structural element that underpins the garden's fecundity.

A second characteristic of the system was the modification of the weather through the use of 'hot beds' and glass. Hot beds are garden beds formed with fermenting manure, cased with dirt, and topped with glass cloches or cold frames.[20] Hot beds might be a few inches high for a fast crop, like radishes, or feet high to start the melons or force the asparagus. They were usually used in late fall and winter, but could also be used in a cold August to help starts. Warming the ground is a season-defying gardening method that enabled high-status tables to offer asparagus in November, and salad greens in January. Extending seasons by modifying temperature was a second important technology for this gardening system. I don't think it is possible to over-emphasize the labor required to maintain hot beds. The glass frames and cloches had to be adjusted as the weather changed. They were opened, closed, and even shaded, depending on conditions.

Gardeners extended the growing seasons two other ways, through the use of different cultivars – early and late maturing varieties – and especially through succession planting. It was with succession planting that one sees the genius of the system and the astonishing productivity of the garden. Succession planting could be as aggressive as lettuce planted every five days to ensure a steady supply of baby greens and, once thinned, mature lettuce plants always in a state of peak perfection.

Martha Bradley, author of *The British Housewife: or, the Cook, Housekeeper's and Gardiner's Companion* (1756), made explicit in her title what is implicit in all early-modern cookbooks. The kitchen cannot be understood without the garden. The following quotation lists the 'Garden Stuff' Bradley said a good gardener should be able to provide in the month of February. I quote her almost in full because she so eloquently suggests a dialogue between cook and gardener in the active management of the garden for the kitchen.

> The Greens in Season in *February* are Cabbage, Savoys, Baccoli[?], and Brocoli, Coleworts,[21] Spinage, and fine Sprouts.[22] Small Salleting also comes in Abundance, especially if the Weather be a little mild, or may be raised at any Rate with proper care;[23] and the Beds, which have been properly managed, yield Mushrooms in sufficient Plenty.
>
> Endive also and Celri are in great Perfection; and there is Chervil… The Asparagus of this Month depends upon the Season for it's Goodness; it is best on the Hot-beds made in December, and if there be a little Sun to colour it, will

But, Did the English Eat Their Vegetables?

be very good, otherwise, it is but sickly.

Among the smaller Herbs there are Parsley, Sorrel, Thyme, Winter Savoury, and Pot Marjoram.

The Roots in Season in *February* are Carrots, Parsnips, and Turnips, Beets and Skirrets, Salsify and Scorzonera, Jerusalem Artichokes, and Potatoes; also Onions, Leeks, Garlick, Shallots and Rocombole.[24]

That is thirty-two vegetables, excluding the unnamed small salletings, which could easily include a dozen different greens in and of themselves. All this in England in February! And all growing just outside the kitchen door! Consider the limited range of vegetables for sale in February in an English country farmers' market, or even in a California farmers' market.

I've mentioned considering the money and material resources invested in the kitchen garden as signs of the importance of vegetables in early-modern cookery, but I think we need to add something that would be outside of an economic analysis of the garden: imagination. There was focus, drive, and many dreams of a winter's night behind Martha Bradley's list of vegetables and the operation of the kitchen garden. For the Adams and Eves of these gardens there must often have been a passion for the garden – and a love for vegetables.

I have been surprised in my own garden how much better certain vegetables I have grown tasted compared with ones purchased even at a farmers' market where they were picked the evening before, or the morning of the market. Given the incredible fertility of the soil in these kitchen gardens – and the system was, in practice, more subtle and complex than just planting everything in horse manure – one needs to imagine intensely flavored vegetables, with the exception, of course, of sickly asparagus grown in hot beds during a sunless January.

Whenever money is lavishly spent there is often an aspect of social display, and thus competitive display. The kitchen gardens existed within a social context. I think one can find a display of garden competitiveness in the fall planting schedule for beans and peas in John Abercrombie's *Every Man his Own Gardener* (1791). I read the schedule as a valorous (and expensive) attempt to have the garden with the first harvest of spring beans and peas. These first peas and beans will, of course, taste delicious served with sweet spring-grass butter. But I think it is also implicit that the taste will be all the sweeter if one can serve them to one's good friend whose beans and peas are, unfortunately for the friend, a week or two behind yours.

Cooking greens, best practice

Most people spoil Garden Things by over boiling them: All Things that are Green should have a little Crispness, for if they are over boil'd they neither have an Sweetness or Beauty.

Hannah Glasse, 1746 [25]

But, Did the English Eat Their Vegetables?

Take one large or two ſmall cauliflowers, waſh them very clean, half boil them, and pull them into ſprigs...

Richard Briggs, 1794

Spinage may be boiled like other Greens, but the very beſt method is in its own Juice with a little salt...

Martha Bradley, 1756

The simple English preparation of boiled vegetables with butter can be an awful dish or a brilliant dish, depending on the quality of the vegetables, the care with which they are boiled, and the quality of the butter. The written record seems to be clear. They started with perfect vegetables, boiled them to perfection, and as a rule simply buttered them with butter from pastured cows – a butter that was, at least during the months when cows were on pasture, also a butter of perfection.

While there are few general guidelines offered for cooking vegetables in early-modern cookbooks – Hannah Glasse's instruction, above, is an exception – there are a number of recipes by many different authors from which one can infer best practice. An unusually comprehensive set of instructions for boiling vegetables is found in Martha Bradley's *The British Housewife*.[26] Her instruction for cooking spinach in nothing more than the water left on the leaves after washing, plus a little salt, says it all.[27] Cook to release flavor, not to lose it. Bradley was a wonderful cookbook writer. She offers the kind of fine detail that helps readers replicate the spirit, as well as the form, of the recipe. In case there is any doubt as to what degree of tenderness she means by a 'tender' string bean, she says,

> And when the Water boils put in the Beans; when they have boiled a little while take one out and taſte it; as ſoon as they are tender throw them into the Cullander…

Control over the degree of vegetable tenderness, and a preference for stopping far short of mush, is implied by Robert Briggs' instruction to 'half boil' a vegetable that was then to be dipped in batter and fried. His instruction is consistent with a desire to retain both the vegetable's inherent flavor and a degree of firmness. An ideal seems to have been a boiled vegetable that was tender, but not mushy, and that retained its color. The crispness that Hannah Glasse mentioned in her instruction, quoted above, was more French than English. As Hannah Glasse was an Anglophile, it may simply have been her intention to suggest to cooks to err on the side of under boiling rather than over boiling.

While Hannah Glasse seemed to have assumed that readers needed instruction on cooking vegetables lest they overcook them, writers for more affluent readers, such as William Verral, assumed that readers shared the ideal of tender, but green. Of asparagus, he wrote that

But, Did the English Eat Their Vegetables?

> The French prefer a crispness and yellow in asparagus and French beans, to what we are always in so much care to make green and tender; but they eat it (as they do so many other vegetable) for a hot sallet.[27]

His instruction for cooking asparagus in the French manner is, thus, to 'boil your grass but a little,' but I think it important to note that his description of English practice – *we are always in so much care to make green and tender* – implies the kind of precise intentionality that one would expect of cooks whose aim is to allow vegetables picked in peak condition to speak for themselves in comparatively simple preparations. It is difficult to know what were standard cooking procedures, but perhaps Verral's 'so much care' reference is to a standard procedure in which the cook routinely tested green vegetables for just tender.

Verral, on string beans, describes the boiling practice that is considered best practice today – plenty of water[28] – and again refers to an English taste for a tender vegetable, but one that is still green:

> For this the French cut their beans as thin as possible, and boil as we do in a vast deal of water, with salt, to preserve their greenness, but not so tender, strain them off, and put 'em to a small ladle of…..

The importance of retaining greenness is implied by the work of other authors. E. Smith wrote, to 'garnifh the difh with coleworts and fpinach fcalded green,'[29] and Richard Bradley, using similar language, instructed that 'Lettuce, Endive, Spinage, or what Herbs you please, [be] boiled green.'[30]

Conclusion

Beautiful vegetables picked the day they were eaten, cooked to perfection, and eaten for pleasure, for health, and for poetry – who could doubt the English ate their vegetables? Early-modern kitchen gardens offer a key to period cookbooks, but they also offer an invaluable source of expertise for farmers today who are trying to provide locally grown produce to a discriminating urban market.

Notes

1. Switzer (1731) associates broccoli, cardoon, and sweet fennel with high-status households and the latest culinary rage.
2. Parkinson (1629), Switzer (1731), Tournefort (1716).
3. Tournefort (1716: v. II p. 53) Asparagus was thought good for urinary problems, specifically to attenuate kidney stones.
4. Op. cit.

But, Did the English Eat Their Vegetables?

5. In the rare example of apparently comprehensive menus, such as in the *The Ochtertyre House Booke* (1737), it is possible to find a way around the problem. In this case, the menus suggest that between May 5 and August 5, 1737, apparently the months when there was produce for salads in the Scottish kitchen garden, a salad of some kind – not necessarily a lettuce salad – was served on 55 of the ninety days, thus 60 per cent of the time. Given the ability of more southerly English gardens to produce greens every month of the year, I think one should assume a substantial number of meals included green salads.
6. I counted herbs and vegetables in the New Leaf Community Market in Santa Cruz, California, on May 28, 2008. I did not count multiple cultivars of a given green, e.g. different varieties of kale or potato. Neither early-modern cookbooks nor garden calendars referred to cultivars, but multiple cultivars existed for almost everything grown. The herbal by Tournefort is particularly useful in understanding the choices available to kitchen gardeners at the beginning of the eighteenth century.
7. See any of the gardening calendars such as Evelyn's *Kalendarium Hortense* (1699). Online editions of Abercrombie's *Every Man his Own Gardener*, though late eighteenth century, is easily accessible through Google Books and is typical of the genre. I recommend the 1787 edition: http://books.google.com/books?id=OuA1AAAAMAAJ&printsec=frontcover&dq=Every+Man+his+Own+Gardener.
8. Switzer (1731: p. 4).
9. Including in those published by Hannah Glasse (1747) and Martha Bradley (1756).
10. See skirret entry in Evelyn's *Acetaria* (1699) and Parkinson (1729).
11. Bonnefons (1658) see melons.
12. Abercrombie (1791: p. 241).
13. Thick (1998: p. 100).
14. Jeavons (1974) and Seymour (1978) offer two late twentieth-century interpretations of what is often called 'French intensive' gardening. They support my interpretation of the early-modern gardening systems as having been unusually productive.
15. Mortimer (1716) entry for carrots.
16. Gardiner (1599: n.p.) gave specific measurements for raised beds which conform to the general principle that the gardener could reach to the center of the bed without stepping on it. Mortimer does not discuss manuring practices.
17. Bonnefons (1656) see first chapter. This 'double digging' system was reintroduced to Anglo-American kitchen gardens in the 1980s. Seymour (1978) was a leading British proponent of the system, and Jeavons, via the British Alan Chadwick, the leading American proponent.
18. Bonnefons (1658) see asparagus.
19. Ibid. p. 23.
20. Ibid. p. 21.
21. This could include what we call kale. They had several varieties.
22. Probably side shoots from bolting Savoy cabbage.
23. Small salletings were what we would call micro-greens. Amongst others, they included lettuce, mustard, cress, radish, chervil, purslane, mint, and sage.
24. Bradley (1758 reprint 1998: p. 128).
25. Glasse (1847 reprint 2004: 18).
26. Bradley (1756 reprint 1998: 156–161).
27. Verrall (1759 reprint 1988: 116) There are many recipes in this book that offer hints on the ideal degree of cooking, as well as what the standard English practices were. For example, see the *Sherdoons* recipe on page 119 for a reference to the standard English practice of simply boiling them and serving them with butter.
28. McGee (2004: 285).
29. Smith (1742: 5).
30. Bradley (1762: 130).

Bibliography

Abercrombie, John. *Every Man his Own Gardener*. London: J. F. & C. Rivington, 1791.

Bonnefons, Nicolas de. *The French Gardiner*. Translated by John Evelyn. 4th ed. London: T. B. for B. Took, 1658.

Bradley, Martha. *The British Housewife, or, the Cook, Housekeeper's and Gardiner's Companion* (1756). Totnes: Prospect Books, 1998.

Bradley, Richard F. R. S. *The Country Housewife and Lady's Director, Etc*. London, 1736.

Briggs, Richard. *The English Art of Cookery*. Dublin: P. Byrne, 1798.

Colville, James, ed. *Ochtertyre. House Booke of Accomps, 1737–1739*. Scottish History Society, vol. 55, 1907.

Evelyn, John. *Kalendarium Hortense: Or the Gard'ners Almanac*. London: Robert Scott, Richard Chiswll, George Sawbridge, and Benjamin Tooke, 1699.

Evelyn, John, and Christopher Driver. *Acetaria: A Discourse of Sallets* (1699). Totnes: Prospect Books, 1996.

Gardiner, Richard. *Profitable Instructions for the Manuring Sowing, and Planting of Kitchen Gardens*. London: Edward White, 1599.

Glasse, Hannah. *'First Catch Your Hare…': The Art of Cookery Made Plain and Easy* (1746). Totnes: Prospect Books, 2004.

Jeavons, John. *How to Grow More Vegetables Than You Ever Thought Possible on Less Land Than You Can Imagine*. Palo Alto: Ecology Action of the Midpeninsula, 1974.

Langley, Batty. *New Principles of Gardening*. London: A. Bettesworth and J. Battey, 1728.

McGee, Harold. *On Food and Cooking: The Science and Lore of the Kitchen*. Completely rev. and updated ed. New York: Scribner, 2004.

Mortimer, John F. R. S. *The Whole Art of Husbandry*. Fourth ed. London: R. Robinson; G. Mortlock, 1716.

Parkinson, John. *Paradisi in Sole Paradisus Terrestris* (1729). Amsterdam and Norwood, N.J.: Theatrum Orbis Terrarum ; W. J. Johnson, 1975.

Seymour, John. *The Self-Sufficient Gardener*. London: Faber, 1978.

Smith, E. Cook. *The Compleat Housewife*. Eleventh ed. London: J. & H. Pemberton, 1742.

Switzer, Stephen. *A Compendious Method for Raising Italian Brocoli, Cardoon, Celeriac, and Other Foreign Kitchen Vegetables*. Fifth edition. ed. London: Thomas Astley, 1731.

Thick, Malcolm. *The Neat House Gardens: Early Market Gardening around London*. Totnes: Prospect Books, 1998.

Tournefort, Joseph Pitton de. *The Compleat Herbal: Or, the Botanical Institutions of Mr. Tournefort Vol. I. London*: printed by John Nutt; sold by J. Morphew, 1716.

Verrall, William, *William Verrall's Cookery Book: 1759*. Irthlingborough: Southover, 1988.

Who Put the Leeks in Cock-a-leekie Soup?

*Allyson E. Sgro**

Leeks were first brought to Scotland by medieval monks along with numerous other vegetables,[1] but only these alliums have been immortalized in one of Scottish cuisine's best-known dishes: cock-a-leekie soup. At its simplest, cock-a-leekie is simply a chicken stock-based soup accented with the texture and taste of leeks. Variations today include adding cream, rice, potatoes, barley, or the 'traditional' prunes to the leeks and chicken broth. While it is considered common knowledge that cock-a-leekie has roots in medieval times, its earliest known ancestor is described in Fynes Moryson's account of the food he encountered during his 1598 tour of Scotland:

> Touching their diet: They eate much red Colewort and Cabbage, but little fresh meate, using to salt their Mutton and Geese, which made me more wonder, that they used to eate Beefe without salting. The Gentlemen reckon their revenewes, not by rents of monie, but by chauldrons of victuals, and keepe many people in their Families, yet living most on Corne[2] and Rootes, not spending any great quantity on flesh. My selfe was at a Knights House, who had many servants to attend him, that brought in his meate with their heads covered with blew caps, the Table being more then halfe furnished with great platters of porredge, each having a little pence of sodden meate : And when the Table was served, the servants did sit downe with us, but the upper messe in steede of porredge, had a Pullet with some prunes in the broth.[3]

Contained in this paragraph is an accurate portrait of the state of the late sixteenth century diet of the Lowland Scots. Period accounts of the lower-class diet, such as Bishop Leslie's in 1568,[4] report the main source of nutrition as oatmeal, supplemented by small amounts of meat, fish, dairy, and such hearty vegetables as kale. By contrast, the diet of the wealthy mainly consisted of flesh. Household accounts and inn bills of fare for the well-to-do detail legs of beef, mutton, poultry, and occasional rabbit and duck appearing on dinner and supper tables. Alongside the flesh appeared wine, ale, and loaves of white and oat bread.[5] Except as an ingredient in broths and pottages, vegetables appear to be entirely absent from upper class tables. How a cuisine almost entirely void of vegetable use could come to celebrate leeks in one of its classic recipes is difficult to imagine.

---

* This paper won the Cherwell Food History Studentship Award, 2008.

Who Put the Leeks in Cock-a-leekie Soup?

Vegetables in the Stuart era

What makes the apparent absence of such vegetables as leeks in late sixteenth- and early seventeenth-century Scottish cuisine most puzzling is the abundance of produce cultivated in the country at the time. Late sixteenth-century travelers report gardens were common on upper class estates with leeks, onions, and kale figuring largely in the seed purchases.[6] However, none of the accounts of vegetable cultivation is detailed enough to give further insight into the full extent of these crops and their uses at the beginning of the seventeenth century.

Adding to the mystery is the evolution of cuisine throughout Europe to give vegetables a more prominent role on the dinner table. Contemporary cookbooks from other nations, such as La Varenne's 1651 *Le cuisinier françois*, have recipes for preparing vegetables also known to have been grown in Scotland at this time and suggestions for integrating these dishes into meals. Cock-a-leekie soup could easily masquerade as French innovation alongside the pottages La Varenne describes. While no extant Scottish cookbooks exist prior to the late seventeenth century, the accounts that remain make it clear that vegetables seldom make an appearance on Scottish plates as the centerpiece of a dish as they do in other countries' cuisines.

The vegetable use we have evidence for during most of the following century is fairly limited and appears to be confined to the poor and middle classes. 'The Blythesome Bridal,' an early seventeenth-century folk song, describes a middle-class wedding feast where spring onions, radishes, kale, peas, and seaweeds figure prominently in the meal.[7] Kale is again mentioned by naturalist John Ray during his visit to Scotland in the 1660s, where he notes the natives 'use much pottage made of coal-wort'[8] but fails to mention any other vegetable use in Scotland. Soups and vegetables are entirely missing from the tables of the wealthy that hosted foreign guests, or left so little impression in comparison to the fish, fowl, and flesh that was served that they were not worth mentioning.[9] It is not until the late 1600s that a thorough account of what Scotland's soil could support was published and estate records confirmed that such a wide variety of produce was regularly cultivated.

Gardening in seventeenth-century Scotland

John Reid, a third-generation gardener, published *The Scots Gard'ner* in 1683. Contained within are directions for cultivating and preparing an abundance of fruits and vegetables. The book's recipe appendix includes multiple uses for an extensive list of vegetables Scottish soil can support: leeks, cucumbers, artichokes, asparagus, beans, peas, lettuces, cabbages, endive, skirrets, onions, cauliflowers, carrots, beets, potatoes, and spinach.[10] Among these recipes is the first explicit mention of pairing leeks with poultry beyond its inclusion as a broth flavoring. Reid says 'you may stove leeks with a cock' before going on to describe the use of onions with meat and geese.[11] Perhaps this suggestion was inspired by some parent of cock-a-leekie soup, although as Reid has no recommendations for meat and fruit combinations it is certainly a few generations removed from Moryson's pullet and prune pottage.

One aspect of Reid's recipes stands out from the contents of other period recipe collections from Scotland; these recipes are very simple in comparison to the collections of pies and desserts for special occasions detailed in other cookery books, focusing on everyday dishes such as pickles and accompaniments for roasts. Scottish household recipe books from the late seventeenth and early eighteenth centuries primarily contain elaborate desserts and other sweets, along with meat pies and roasts flavored with berries and expensive spices. With bread-thickened sauces and heavy spices such as nutmeg and mace, the recipes are more reminiscent of medieval cuisine than the new culinary tradition ushered in by France's La Varenne a half century earlier.

Scottish vegetables in use

While visitors to Scotland still did not consider vegetable gardens worth mentioning even when discussing crops and other foodstuffs,[12] we do have evidence that these very same vegetables that are almost absent from recipe books are being cultivated and moved from family estates with farmland to other establishments. Pinkie House records of fruits and vegetable sent to Edinburgh and Yester Castle from 1680 to 1705 show artichokes, asparagus, leeks, cauliflower, cabbage, spinach, and a plethora of savory herbs being cultivated.[13] Bills for provisions at Yester show that in addition to these deliveries from Pinkie, additional cabbage, turnips, carrots, celery, beets, shallots, and asparagus were being purchased for the estate.[14] But the question still remains: what were these vegetables used for outside of broths?

There are hints of vegetable consumption in dinner menus as side dishes for the copious amounts of meat gracing the table. A menu served at Dunrobin Castle on 2 November 1703 reads:[15]

> Brown Soup, 2 hochs, & one piece of Beef
> Boiled mutton and Cauliflowers, one hind quarter
> Almond Pudding
> Cod's head
> Haggis and Puddings
> ───────────
> Roast Turkeys for relief, 2
> ───────────
> Roast Rabbits, 4
> Salad
> Roast Beef, one back 'sye'
> Sheep's-head Collops, 4
> A 'manchett'
> One Roast Plover for my Lady
> One piece of Beef, and a flank for the Servants

Cauliflower and a salad that may have contained greens or other vegetables hardly encompass the range of vegetables grown in Scotland at the time. However, while the modern eye may be accustomed to seeing a wide variety of produce available year-round, before the advent of refrigerated shipping all produce was served fresh in season or preserved. Once this is taken into account, the amount of vegetable consumption at this time seems entirely reasonable. Menus at Dunrobin from January through March list 'roots' as the only vegetable served on a regular basis. In the spring and summer months, dinners and suppers feature spinach, with artichokes added into the mix in July. By September, beans have replaced the artichokes and later in the month we begin to see a shift back toward root vegetables.[16] While there's no cock-a-leekie or even a 'Leek Soup, quarter pullet' listed in any of the menus, it would not be out of place on the table at Dubrobin. A dish served at dinner on 9 May 1704 is a close relative of Moryson's broth and today's cock-a-leekie: 'Selery Soup, quarter poullett.'[17]

Household recipe books from similar homes give us a closer look at other ways this produce might be utilized in wealthy homes. Roasts, soups, and pies detailed in household manuscripts contain not only the fruits and spices of medieval-style cuisine, but also integrate the vegetables being grown in Scotland. In one personal book marked with the initials M.I.M. from the late seventeenth century, there is a soup recipe for chicken, veal, or lamb, with spinach and sorrel.[18] Katherine Bruce's 1688 recipe book gives directions for a dish of boiled artichoke bottoms with cream, ginger, and sugar.[19] A late seventeenth-century recipe for chicken pie straddles the two styles with lemon, artichokes, cinnamon, barberries, dates, raisins, apples, eggs, and sugar in addition to the chicken.[20]

So where are the vegetables?

Clearly it is not vegetables that are absent from early eighteenth century Scottish cuisine, but the notion of vegetables being the focal point of a dish or meal. It is still flesh that is celebrated at the table and considered the centerpiece of the meal. Geddes suggests this may be due to vegetables being regarded as a delicacy and thus unsuitable as a main course.[21] However, vegetables are being served multiple times in a meal on a nightly basis in households such as Dunrobin's, are served in inns throughout the Lowlands,[22] and have been an integral part of soups in Scotland for centuries. It seems unlikely that all vegetables would be regarded as such a rarity when they were served in a variety of venues by all classes. Furthermore, if vegetables were such a delicacy, we would expect to see a wide variety of recipes featuring their use in the special-occasion cookery books of the time beyond the handful of artichoke dishes,[23] vegetable-flavored creams, and pickle recipes. Instead, they are still mainly used to accent some other primary ingredient.

During the first few decades of the 1700s, both the number of recipe books available to us from the era and the frequency of vegetable recipes increases dramatically. Lady Castlehill's personal recipes from 1712 contain pea soup and artichoke broth[24] and a 1709 collection has directions for a pea tart.[25] To ensure utilization of vegetables year-round, there are pickling directions for mushrooms, beans, cucumbers, and cauliflowers.[26] Even in Mrs McLintock's 1736 published cookery book that focuses mainly on pastry work, the short soup section has recipes for onion, pea, and greens soups.[27] While books still tend toward meat and pasty recipes, this is a notable change from a few decades before when only a handful of artichoke recipes and vegetable-flavored creams[28] were recorded. It is in this rapidly changing culinary scene that cock-a-leekie soup is finally mentioned by name for the first time.

Cock-a-leekie appears by name

In Ochtertyre House's account books, the amounts of food bought and the food served are recorded starting in January, 1737. For dinner on 19 January 1737, the first dish recorded is 'Cockie Leekie fowls in it.'[29] Throughout the winter and spring months, 'cockie leekie' is served, as well as green broth, 'celery soop ducks in it', and cabbage broth. After 24 April, cock-a-leekie is not served again until the following January, but throughout the year a wide variety of other vegetables graced the table at Ochtertyre House: asparagus, celery, leeks, salad greens, cabbage, kail, artichokes, spinach, potatoes, peas, and beans. In the summer at least one vegetable is served a day, if not multiple varieties.[30] Vegetables are finally fully integrated into daily Scottish cuisine.

What the emergence of cock-a-leekie soup indicates is not that vegetables were being utilized for the first time, as there is evidence that vegetables were in use for decades before the 1730s. However, only at this time were they accepted as ingredients worthy of showcasing as part of the main course and remembering in recipe titles. So who put the leeks in cock-a-leekie soup? The Scots did, putting to rest the long-standing notion that they have traditionally not enjoyed vegetables and laying the foundation for modern Scottish cuisine.

Cock-a-leekie Soup Recipe

Before I wrote this paper, I must confess that I'd never tried cock-a-leekie soup. As a lover of both chicken soups and leeks I wasn't disappointed. The prunes add an unexpected richness to the simple chicken stock that complements the sweetness of the leeks. This recipe was my favorite variation.

1 chicken (1–2 kg)
4 leeks (about 1 kg)
6 carrots
½ onion
20 prunes, stoned
6 quarts water
salt and pepper to taste

Coarsely chop the carrots, onions, and green part of the leeks. In a stock pot, combine the chicken, water, carrots, onion, prunes, and the green portion of the leeks. Bring to a boil, cover and let simmer for an hour, and then allow the stock to cool.

 Remove the chicken and vegetables from the pot, cut the chicken off the bones and chop into small pieces. Chop the whites of the leeks into small pieces, add them and the chicken to the pot, and bring it back to a boil. Let simmer for 20 minutes, salt and pepper to taste, and serve.

Who Put the Leeks in Cock-a-leekie Soup?

Notes

1. Dickson, p 29.
2. Davidson, p. 217. Corn in this context does not refer to maize, but to cereal crops in general. Scotland's major cereal crops of the time were oats and barley.
3. Brown (1891), p. 88.
4. Brown (1893), p. 167. Bishop Leslie described the diet of those in the Borders region as, 'fleshe, milk, and cheis, and sodne beir'.
5. Geddes, pp. 4–11.
6. Hope, p. 163.
7. Johnson, pp. 58–59.
8. Brown (1891), p. 231.
9. Brown (1891), pp. 126–127.
10. Reid, pp. 115–120.
11. Reid, p. 119.
12. Brown (1891), pp. 272–275. Thomas Morer visited Scotland in 1689 and later produced a detailed account of his travels where he described the customs of both the Highlands and Lowlands in great detail. His discussion of Lowland food covers their grain crops, bread, meat consumption, cheese, poultry, and fruit orchards. Not once is the cultivation or consumption of vegetables mentioned.
13. Pinkie Household Papers, ff. 9, 49, 70–72.
14. Yester Bills, ff. 249–250, 252, 278.
15. Dinner and Supper Bills of Fare at Dunrobin, November 2, 1703.
16. Dinner and Supper Bills of Fare at Dunrobin.
17. Dinner and Supper Bills of Fare at Dunrobin, May 10, 1704.
18. M.I.M., ff. 73, 186.
19. Bruce and Unknown, f. 4.
20. Cookery and Medical Recipes, f. 6.
21. Geddes, p. 26.
22. Inn, dinner and horse bills, ff. 99, 101, 116–119; Pinkie Household Papers, ff. 77, 89, 95.
23. Maule, ff. 1, 9, Lockhart, p. 10, McLintock, p. 7. Janet Maule gives recipes for an artichoke pie and fried artichokes, while Lady Castlehill describes an artichoke broth. Artichokes also appear regularly in meat pies, such as the ones described by Maule and McLintock.
24. Lockhart, pp. 10, 52.
25. Recipe Book (1709), f. 5.
26. Lockhart, pp. 28, 44, McLintock pp. 41–44.
27. McLintock pp. 47–49.
28. Smith, f. 5, Maule, ff. 1, 9, 26.
29. Ochtertyre Household Account Book, f. 4.
30. Colville, pp. 1–245.

Bibliography

Manuscripts

'Accounts of Household and Personal Expenses and Accounts of Fruit and Vegetables Sent to Edinburgh and Yester, 1662–1778.' In *Pinkie Household and Estate*. Edinburgh: National Library of Scotland.

Bruce, Katherine, and Unknown. 'Recipe Book, 1688.' In *Fletcher of Saltoun Papers: Household*. Edinburgh: National Library of Scotland.

'Cookery and Medical Recipes, Late seventeenth Century to 1734.' In *Macadam Collection*. Edinburgh: National Library of Scotland.

'Dinner and Supper Bills of Fare at Dunrobin.' In *Sutherland Estates Papers*. Edinburgh: National Library of Scotland.

'Household Account Book for Ochtertyre and Fowlis Easter, 1737–9, with Daily Menus and Lists of Expenditures.' In *Murray of Ochtertyre Accounts*. Edinburgh: National Library of Scotland.

'Inn, Dinner and Horse Bills, Scotland and England, 1643–94.' In *Personal and Household (Yester, Edinburgh and London) and Yester Estates: lst Marquess of Tweeddale*. Edinburgh: National Library of Scotland.

M.I.M. 'Household Recipes, 17th Century–18th Century.' Edinburgh: National Library of Scotland.

Maule, Janet. 'Recipe Book of Janet Maule' 1701. Edinburgh: National Library of Scotland.

'Recipe Book, in Several Eighteenth-Century Hands.' In *Fletcher of Saltoun Papers: Household*. Edinburgh: National Library of Scotland.

Smith, Kathrin. 'Recipe Book' 1697. In *Recipes and Agriculture*. Edinburgh: National Library of Scotland.

'Yester Estates Bills for Provisions, 1685–6.' In *Yester Papers*. Edinburgh: National Library of Scotland.

Printed material

Brown, P. Hume, ed. *Early Travellers in Scotland*. Edinburgh: David Douglas, 1891.

———, ed. *Scotland before 1700: From Contemporary Documents*. Edinburgh: David Douglas, 1893.

Colville, James, ed. *Ochtertyre. House Booke of Accomps, 1737–1739*. Scottish History Society, vol. 55, 1907.

Davidson, Alan. *The Oxford Companion to Food*. Oxford: Oxford University Press, 1999.

Dickson, Camilla. 'Food, Medicinal and Other Plants from the 15th Century Drains of Paisley Abbey, Scotland' *Vegetation History and Archaeobotany* 5, no. 1–2 (1996): 25–31.

Geddes, Olive M. *The Laird's Kitchen: Three Hundred Years of Food in Scotland*. Edinburgh: Her Majesty's Stationery Office 1994.

Hope, Annette. *A Caledonian Feast*. Edinburgh: Canongate Books, 2002.

Johnson, James. *The Scots Musical Museum: Consisting of Six Hundred Scots Songs with Proper Basses for the Piano Forte*. 6 vols. Vol. 1. Edinburgh: James Johnson Music Seller, 1803.

La Varenne, Francois Pierre. *The French Cook*. London: Charls Adams, 1653.

Lockhart, Martha. *Lady Castlehill's Receipt Book: A Selection of eighteenth Century Scottish Fare: Original Recipes from a Collection Made in 1712*. Edited by Hamish Whyte. Glasgow: Molendinar Press, 1976.

McLintock, Mrs. *Mrs. Mclintock's Receipts for Cookery and Pastry-Work : First Published 1736*. Aberdeen: Aberdeen University Press, 1986.

Reid, John. *The Scots Gard'ner: In Two Parts*. Edinburgh: David Lindsay and Partners, 1683.

Bone-dry Freshness: Dried Vegetables

Aylin Öney Tan and Filiz Hösükoğlu

The word fresh is magical; health and life seem to flow from the word itself. Indeed, a vegetable is fresh when it still retains its water. As water is the source of both life and health, freshness means being full of life, and not having yet lost life.

However, we call a vegetable fresh when its connection with life is actually just severed. We call vegetables fresh when they have been recently picked from their branches. Once picked, time mercilessly flows to erode the freshness of a vegetable, making it slowly 'go bad.' When this lifeline with soil is severed, decay is imminent. The water going to the vegetable is cut from the source; life stops, the process of disintegration speeds up.

Yet precisely at this hour of death, human beings have discovered the secret to a second life: drying vegetables. This is yet another chapter in human beings' incessant struggle against nature. In this way, vegetables, which have lost their water and hence, their life, are given a new lease of life.

Anatolia is home to one of the oldest and most diverse traditions of drying vegetables. Anatolian people, living in a land that is the cradle of civilization, discovered long ago that drying is the ideal way to preserve fresh vegetables and they constitute an important part of Turkish cuisine. The need for vegetables during long winters might have been the original starting point of this tradition but now many dishes are preferably cooked with dried vegetables.

The range of vegetables that are dried is simply amazing. All across Anatolia peppers, eggplants, zucchinis, and marrows are hollowed out and dried for stuffed dishes. Strips of squash, chopped green beans, halved tomatoes, stringed baby okras, husked corns are turned into soups, casseroles, stews. Regional tastes include such dried wild greens as melokhia, Indian knotgrass, etc. This paper offers a glimpse of the world of dried vegetables in Anatolian kitchens. One final example is *tarhana* – a soup made of dried vegetables.

The tradition of drying food

Drying foodstuffs is a necessity that has evolved into a social ritual. The drying process requires quickly preparing huge mounds of vegetables; it is a race against the decaying process. The perfect time for drying usually falls between the last week of July through the first week of September as this is when fresh vegetables are ample and the conditions are most suitable. The old Ottoman calendar used to signify certain dates with relevant winds and heat waves. The perfect time for drying – when the weather is hot and dry

Bone-dry Freshness: Dried Vegetables

– has a name of its own, *eyyam-ı buhur,* which is considered the hottest period of the year, falling between 2–9 August. Nowadays, with global warming, the precise weather calendar of earlier times seems to be changing.

Reflection on architecture

The practice of drying food for winter has had an influence on the spatial articulation of the traditional Turkish house as well. The traditional Turkish house has a stone ground floor with a paved courtyard or vestibule. The ground floor is not for living but for utilities and in some cases for animals. The living quarters are situated in the timber-framed upper floors. In between the ground floor and the first floor there is usually a hidden mezzanine, specially designated for food storage. This semi-hidden section of the house is reached through the staircase connecting the floors. There is usually an elevated platform enlarging the landing, which provides space for food preparation and laying out the drying food. This space is built entirely with timber and it is ventilated by latticework. The food storage area is dry as it is separate from both the humid ground floor and the heated living quarters above. The ventilation holes provide air circulation and keep the food dry and cool. Dried vegetables are often kept hanging on strings or in cotton bags. When family members or neighbours gather to prepare food for winter, this is the spot to work.

Legumes and root vegetables

Legumes are almost always available in dried form. They are not included in this paper as they are not categorized as vegetables in Turkey. Many varieties of chickpeas (*nohut*), white kidney beans (*kuru fasulye*), lentils [yellow, red, and green varieties] (*mercimek*), broad beans (*bakla*), borlotti beans (*barbunya*), black eyed beans (*börülce*), mung beans (*maş*), and corn (white and yellow) are usually available in dried form. Borlotti beans are preferred in fresh form though a dried version is widely available. Fresh chickpeas are only for nibbling as a snack and fresh fava or broad beans in the pod are a particular spring delight for cooking. Apart from these few examples, legumes are consumed dried. Drying legumes does not need any elaborate treatment. However, one particular example for drying lentils is worth mentioning. I have witnessed its preparation in Anatolian province of Afyon in the house of a poppy oil manufacturer. Lentils were mixed with chopped onions, red and green peppers and salted. The resulting mélange resembled an instant meal fix, ready to be transformed into a dish simply by boiling in water.

Traditionally there is no account of drying root vegetables. Root vegetables like carrots, celery root, beets, Jerusalem artichokes, turnips, potatoes, etc. are winter vegetables that were never considered for other seasons. These keep pretty well for a long time and they can be buried in sand or hay, so there is no need to dry them.

Bone-dry Freshness: Dried Vegetables

Classics: eggplants, courgettes, peppers, tomatoes

The need for winter nutrition is an impetus for drying vegetables but there are three vegetables that are preferred in their dried state: eggplants, courgettes, and peppers. These vegetables are dried not for the purpose of preservation alone but because their taste is considered improved when dried. Tomatoes can also be classified with this trio as it adds summer shine and taste to almost all dishes.

Eggplant is definitely the king of the dried vegetable world. When preparing eggplant for drying, not a bit is wasted. The most common method is to hollow it out so it can be used for stuffed dishes. The skin is retained because it helps the eggplant to hold its shape when cooked. The vegetable is halved, the tops cut off, and the insides hollowed out. Eggplant is never salted in the drying process. The tops and flesh are dried and boiled for *musakka* in winter. If the eggplant is going to be used for other dishes it is quartered lengthwise, keeping the stalk intact.

Courgettes and gourds are dried to be used as *dolma* shells. First their ends are chopped off. Their insides are carved out with a tool designed for that purpose. They are slightly salted to release their water. In southeastern and southern provinces (Gaziantep, Urfa, Mersin, Tarsus, and Adana), two major types of the *cucurbitaceae* familiy are dried: Armenian cucumber, tr. *Acur* (*Cucumis melo flexuosus*) and vegetable marrow, tr. *Haylan kabağı* (*Cucurbita pepo*).

When drying peppers the seeds of hot and sweet peppers are removed and the hollowed peppers are dried in the sun. When they're used as a spice, they are ground on a black basalt grinding stone. Salt and a little olive oil are also added for taste and shine.

Tomatoes are cut horizontally, sprinkled with coarse salt, placed on wooden trays, and set in the sun to dehydrate. At 40 degrees Celsius, it takes only a couple of days for the tomatoes to dry properly. The dried tomatoes are kept in cotton bags for longer shelf life. Sun-dried tomatoes are used to flavor soups, stews etc.

Other favourite dried vegetables

Green beans (*Phaseolus vulgaris*)
Tr: *Fasulye, Ayşekadın fasulye, Çalı fasulye*

Drying green beans is a common practice throughout the inner and coastal parts of the Black Sea region. The flat bean variety is used for drying. Green beans are cut diagonally into bite-sized lozenge shape and hung on a string. Dried green beans are briefly boiled, then dried, and cooked the usual way. They can be cooked with olive oil, without meat, to be consumed cold or at room temperature, or cooked with meat and tomato paste, served as a hot main course. Dried green bean sautéed with onion and scrambled with egg is a popular breakfast dish in the Black Sea region. Dried green beans have a particular smoky flavor that goes well with eggs.

Bone-dry Freshness: Dried Vegetables

Okra (*Abelmoschus esculentus*)
Tr: *Bamya*, (*çiçek bamya*/flower bamya is a term used for tiny, bud sized dried okras)
> A gift of Africa to Anatolian cuisine, okra is consumed fresh as well as dried. However, dried okra is preferred for certain dishes. *Bamya çorbası* (okra soup) of Konya is almost always made with dried okra. Tender, tiny okra of Amasya is renowned. Berrin Torolsan has a wonderful article on the historical significance of okra in Turkey. Okra was already quite an established vegetable by the fifteenth century; in fact, the loyal horsemen who supported Mehmet the Conqueror adapted it as the emblem of their tournaments. Torolsan also cites accounts of travelers.

Leaves, wild greens and herbs
> Many greens are dried to be used as spices. However there are also a great variety of wild greens, which are dried to be reconstituted and used as vegetables. Generally the dried wild greens are added to soups especially yoghurt based soups and added to *bulgur* pilaf.
>
> During my survey I have noticed that the wild greens preferred for drying tend to be more mucilaginous. The fibrous quality of such wild greens as purslane, mallow, and melokhia make them keep better in the drying and reconstitution process.

Wild purslane (*Portulaca oleracea*)
Tr: *Semizotu*, *pirpirim* (generally used for the wild variety)
> Wild purslane, very popular in Turkey, is considered a vegetable rather than a salad green, though it is consumed in both ways. Dried wild purslane is cooked in winter with the addition of black eyed beans, lentils, bulgur, dried tomato, garlic, and dried mint.

Mallow (*Malva sylvestris*)
Tr: *Ebegümeci*; local dialect: *Gömeç, kömeç*
> Mallow is often sautéed with onion and, occasionally, a handful of bulgur. However when dried it is more used in soups or added to *bulgur* dishes. The ratio of bulgur to green is much higher in the dried version of the dish.

Knotweed (*Polygonum cognatum*)
Tr: *Madımak*
> Knotweed is the signature plant of the Central Anatolian city of Sivas. There are songs and folk dances dedicated to this much loved spring green. It is also widely consumed in other central provinces such as Ankara, Çorum, and Tokat. It is almost always cooked with *bulgur* (cracked wheat) and *pastırma* (spiced cured beef). The dried *madımak* is cooked the same way or added to yoghurt-based soups.

Melokhia (*Corchorus olitorius*)
Tr: *Mülhiye, molehiya*
> Melokhia is known only in the southern provinces because such port cities as

Mersin, Antalya, and Alanya, which were in close contact with Egypt, Africa, and Cyprus. There is a considerable population of African origin and melokhia with lamb or chicken is their favourite dish. Dried melokhia is never ground, nor is it made into a soup, but always cooked as a vegetable with meat or chicken.

Nettle (*Urtica dioica*)
Tr: *Isırgan*; local dialect: *Dalan*

Nettle is a widely used green in Turkey. Fresh nettle is usually sautéed with onion or used as a filling for *borek,* savoury pies. However, the dried version is used for yoghurt-based soups or as a tea or infusion.

Smilax
Tr: *Silcan*; local dialect: *Dikenucu, melucan*

This wild green is a favourite of Kastamonu and other northern provinces. As in green beans it is reconstituted in boiling water, drained, and scrambled with sautéed onions and eggs.

Arum
Tr: *Nivik, Yılan yastığı*

This wild green is poisonous if not treated properly. When dried it loses its toxic property and can be safely eaten.

Wild lily shoots (*Asphodelus aestivus*)
Tr: *Çiriş*

Young, wild lily shoots are another favourite of such eastern provinces as Kars, Van, and Erzurum as well as the southern Taurus mountains. The joy of eating this tasty green is extended by drying the chopped shoots. Both the fresh and dried are scrambled with sautéed onions and eggs, added to bulgur dishes, cooked in milk, added to soups, or used as *borek* filling.

Dried herbs

Although dried herbs should not be classified under the category of dried vegetables, they are briefly worth mentioning. No Turkish kitchen is complete without dried mint and thyme. The most common dried herbs are mint, thyme, purple basil, tarragon, parsley, and dill.

Mint (*Mentha aquatica*)
Tr: *Nane* (Dried Mint: *Kuru nane*)

Dried mint is the final touch to almost all Anatolian soups. Mint leaves are separated from the stalks, laid on cloth or newspaper, and dried in the shade. Dried mint is generally crushed into flakes or powder. In some regions the leaves are left whole so as to keep the flavour strong.

Bone-dry Freshness: Dried Vegetables

Tarragon (*Artemisia dracunculus*)
Tr: *Tarhun, tahrin*

Tarragon is used in only four provinces: Ankara, Erzurum, Urfa, and Gaziantep. Fresh tarragon is used as a salad green but the dried version is a must for soups, lentil *kofte, borek* filling, and lentil soup. It is also a preferred herb in preparing *tarhana*. It is predominant particularly in Gaziantep cuisine.

Parsley (*Petroselinum crispum*) Tr: *Maydanoz*;
Dill (*Anethum graveolens*) Tr: *Dereotu*

Parsley and dill are the un-identical twins of Turkish herbs. Nowadays, they are available fresh all year round though both were used dried in the Ottoman court. Dried parsley and dill are not common anymore and generally not commercially available.

Purple basil (*Ocimum basilicum*)
Tr: *Reyhan*

Purple basil is another herb usually preferred in the dried form. It is mostly popular in southern provinces giving flavour particularly to stuffed dishes.

Pennyroyal (*Mentha pulegium*)
Tr: *Yarpuz, narpuz, yabani nane, filiskin, pülüskün*

This herb is considered a must in hot or cold yoghurt soups of the eastern provinces of Erzurum, Erzincan, Elazığ, Kars, and Van.

Coriander (*Coriandrum sativum*)
Tr: *Kişniş, aşotu*

Like pennyroyal, dried coriander is added to *tarhana* soup; fresh coriander is also used in the preparation of *tarhana*.

Tarhana: an archaic soup mix

Tarhana is originally a grain and yoghurt dough, fermented and dried in pellets. It is a very early example of a dried soup mix, reconstituted to make a porridge-like soup. The basic *tarhana* mix is rather simple and bland, with the plain flavour of grain and tangy zest of fermented yoghurt. However, *tarhana* has greatly evolved in Anatolia over centuries. There are as many *tarhana* variations as there are provinces or towns. It may be ground, dried in sheets, dried in cake forms or preserved in jars as a paste. The greatest twist to the original soup mix was the introduction of peppers and tomatoes to Anatolia. The original grain-based recipe gradually evolved into something completely different. The sweet brightness of sun-ripened tomatoes, the tang of paprika peppers, the grassy taste of green peppers, and the pungent aroma of onions and garlic became predominant. Herbs of all kinds also started to appear in the recipes. Nowadays the starter *tarhana* dough is in many cases a great mix of vegetables and herbs. However, one particular herb is essential: *tarhana otu*, or more scientifically *Echinophora sibthorpiana*.

Bone-dry Freshness: Dried Vegetables

This particular herb not only adds flavour but aids the fermentation process of the *tarhana* dough and improves its keeping qualities.

Sustainable method

Drying is the oldest and easiest method of preserving food and drying vegetables. Like many things old and easy, drying vegetables is an example of human ingenuity. The vegetables are healthy as well as ecologically friendly. They bear none of the health hazards associated with chemical additives in processed foods. While it is easy to appreciate fresh vegetables when they are fresh, drying them allows us to enjoy them longer without consuming the energy required for refrigeration or freezing. Thus, in this age when we are ever more conscious of what we choose to eat, drying vegetables is a tradition definitely worth saving.

Select bibliography

Baytop, Prof. Dr. Turhan. *Türkçe Bitki Adları Sözlüğü* [A Dictionary Of Vernacular Names Of Wild Plants Of Turkey]. Atatürk Kültür. Dil, ve Tarih Yüksek Kurumu, 1994.

Bilgin, Arif. *Osmanlı Saray Mutfağı/1453–1650,* [Ottoman Court Cuisine/1453–1650]. İstanbul: Kitabevi, 2004.

Değirmencioğlu, Nurcan, D. Göçmen, A. Dağdelen, F. Dağdelen. 'Influence of Tarhana Herb (*Echinophora sibthorpiana*) in Fermentation of Tarhana, Turkish Traditional Fermented Food.' *Food Technology and Biotechnology*. 2005, 43(2), 175–179.

Fahriye, Ayşe. *Ev Kadını [Housewife]*. İstanbul: Mahmud Bey Press, 1882–1883; reprint: İstanbul: Ofset Yapımevi, 2002.

Halıcı, Nevin. *Konya'da Kışlık Yiyecekler Üzerine bir Araştırma* [A Research on Winter Food Preparations in Konya]. Konya, 2000.

Özcan M., A. Akgül. 'Essential oil composition of Turkish pickling herb (*Echinophora tenuifolia* L. subsp. *sibthorpiana* (Guss.) Tutin),' *Acta Botanica Hungarica*. 45:1–2:163–167, May 2003.

Özsabuncuoğlu, Özden. *Dört Mevsim Gaziantep Yemekleri* [Four Seasons of Gaziantep Cuisine]. Gaziantep: Gaziantep Üniversitesi Vakfı Yayını No 13, 2003.

Tokuz, Gonca. *Gaziantep ve Kilis Mutfak Kültürü*. Gaziantep: Gaziantep Üniversitesi Vakfı Yayını No: 12, 2002.

Torolsan, Berrin. 'Winning Ways with Okra/Championed by the Cavalry.' *Cornucopia*, Issue 34, Volume 6, 2005.

Üçer, üjgân. *Anamın Aşı, Tandırın Başı – Sivas Mutfağı* [Sivas Cuisine – Warmth of Fireplace, Taste of Mother's Food]. İstanbul: Kitabevi, 2006.

Dokonjo Daikon: The Radish with the Fighting Spirit

Michelle Toratani

In July 2005, *dokonjo daikon* (ど根性大根), translated loosely as 'the radish with the fighting spirit' or 'gutsy radish,' made headline news when it was found growing through thick asphalt in Aioi city (相生市), near Kobe, Japan. *Dokonjo* daikon (nicknamed Daichan だいちゃん or little daikon) was seen as an inspirational metaphor for beating the odds during economically depressed times. Rather than removing Daichan from the asphalt to prevent anyone from stepping or tripping on it, the city council erected a sign that asked residents and visitors alike to 'observe him with affection'.[1]

The city of approximately 33,000 inhabitants decided to adopt the daikon and it became a morale-boosting mascot. Daichan was nurtured like a beloved pet and when vandals decided to decapitate the radish one night in November 2005, this heartless act outraged the townspeople, bringing some to tears, and sparked a surprising and bizarre national uproar. There was attention from both national and international media, including morning talk shows. BBC News reported that 'the Japanese public has frequently been touched by the plight of stricken animals,' but commentators were 'at a loss to explain this wave of affection for a mere vegetable.'[2] The Japanese are no strangers to anthropomorphizing animals and food items; they are after all, the land of Hello Kitty. But, articles written particularly in Western countries were done so with a tongue-in-cheek tone. All this media attention probably caused the assailants to regret what they had done and within days Daichan's 'head' was found returned, near its bottom.[3]

Immediately following the return of the 'head,' town council members rushed it to a nearby agricultural research lab and placed it into a shallow dish filled with mineral water in an attempt to revive it. The idea was to save Daichan, by extracting either its DNA or its seeds in order to produce offspring from the original radish.[4] For the new year, Daichan was even displayed in Aioi Town Hall (in its dish) and some 600 people came to visit it, including a group of high school students who used the radish's motivation to thrive through adversity as a sort of good luck charm toward passing Japan's notoriously difficult university entrance exams.[5] By March 2006, 11 cuttings had been taken from the original radish and approximately 400 seeds were collected by the start of 2007. Currently, Daichan's offspring are in Himeji city pending a larger production of seeds intended for sale to the public.[6]

Daichan's amazing tenacity was soon turned into two endearingly illustrated children's books by Ayumi Miyazaki, thus anthropomorphizing, enshrining, and immortalizing Dokonjo Daikon.[7] Sadly, all of this expenditure on Daichan of taxpayer's yen caused the Aioi city council to order town officials and residents to stop talking about Daichan. The concern was that the town would become the target of even greater international ridicule as well as make the residents appear to have too much time on their hands.

The odds of any seedling breaking through asphalt are slim, but the intense outpouring of emotion for the daikon was probably due to its popularity and importance as a vegetable in Japan – it is unlikely a carrot would have had the same impact.

Characteristics of daikon in Japan

Many historians believe that daikon (大根, *Raphanus sativus*) originated from the area between Central Asia and the Mediterranean Sea and made its way to Japan via China and Korea over a thousand years ago. Radishes were used for food in China and Korea as early as 400 BC.[8] Daikon belongs to the *brassica* or cabbage family (including Chinese cabbage and radishes) and are essential to Japanese cuisine and its foodways. There are more than 100 regional varieties of daikon in Japan such as *Nerima*, *Karami*, *Minowase*, *Shōgōin*, *Sakurajima*, *Miura*, and *Hanazukuri*, ranging in shapes from stumpy to slender and colors ranging from the ubiquitous white to rarer colors such as red and deep purple.[9]

Today daikon can be grown year-round, but because the Japanese are seasonal eaters, they instinctively know that the best crop is harvested in late autumn to early winter. Easy to cultivate and preserve, daikon have always been an important crop in Japan, traditionally providing rural populations with their only source of fresh vegetables (along with cabbages) throughout the dead of winter.[10]

Typically in today's market, the most recognizable daikon in markets is the *Aokubi* (green necked) daikon, which is an F1 hybrid created and introduced by a seed company in Japan for maximum yield.[11]

Culinary uses

Every part of the daikon is edible including the tops and peels and is used regularly in Japanese cuisine. It is a shame that when daikon is sold outside of Japan, the dark, slightly hairy green tops are often lopped off as they perish weeks faster than its large, firm root.[12] Elizabeth Andoh's book *Washoku* includes rarely seen English-language recipes for using both daikon peels as well as the green tops in everyday cooking.[13]

Daikon can be eaten raw, cooked, dried, and pickled (dried and pickled daikon will be discussed later). Raw daikon is often sliced into salads or shredded to accompany sashimi; grated daikon can be served as a condiment or flavoring for sauces and is sometimes used to clean oysters to rid them of dirt and impurities.[14] In its raw state, daikon has digestive enzymes similar to natural gastric fluids with its fresh, vibrant, and mustard-like sharpness cutting through to help digest the oiliness in such fish as mackerel, and of other commonly eaten Japanese fish.[15]

Daikon is at its most versatile and transformative when cooked. Unlike many other vegetables, daikon can withstand slow, long cooking techniques such as stewing and will hold its shape well. Whether sliced into miso soup or cooked as part of the hearty winter dish *Oden*, daikon fully absorbs the flavor of the liquid it is cooked in. Cooking also mellows its 'bite' and the juicy meatiness of cooked daikon presents yet another textural dimension.

The versatility of daikon and its textural morphing is probably why it is also an essential ingredient in *shōjin ryōri* (精進料理), the vegetarian Buddhist cuisine that improves one's training and practice of Buddhism through the consumption of only the simplest foods.[16] It is easy to see why daikon would meet both vegetarian requirements, and also the principles of simple cooking: everything from daikon soup stock made from the skins to simmered daikon with miso sauce and braised daikon with deep-fried tofu is made in temples throughout Japan.[17]

Preservation

Tsukemono (pickles) are to Japan's culinary identity as cured meats and cheeses are to Europe.[18] Peter and Joan Martin remark that the 'real Japan' is found in the preserved pickles of which every town will have their own regional product.[19] Therefore, every meal that claims to be authentically Japanese must include pickles to accompany the rice.[20] *Tsukemono* and rice (the two most important components of the Japanese diet) are served at the end of all traditional Japanese banquets as a way of fully sating the diner at the conclusion of a multi-course meal.[21]

As mentioned earlier, the preservation of vegetables was necessary because vegetables could not grow during the coldest winter months.[22] So important is *tsukemono* that in some regions in traditional Japan, a girl was only considered eligible for marriage once she mastered both the technique of cooking rice and art of making pickles perfectly.[23]

Every conceivable vegetable is used for pickling including eggplant, turnip, scallion, ginger, cabbage, and cucumber, but daikon is among the most popular.[24] There are literally dozens of techniques for making pickles, but the most common method of pickling daikon is *takuan,* a pickle that has been made for at least 400 years.[25] *Takuan* is widely believed to have been created by Takuan Oshō, an early seventeenth-century Zen priest who served at Tōkaiji Temple in modern-day Tokyo.[26] Before any pickling begins, the daikon is harvested and hung to dry whole in a sunny, well-ventilated place to remove as much moisture as possible (similar to drying fish or aging meat). The drying time depends on when the pickles are to be eaten – ranging from one week for fairly immediate consumption, to four weeks for later use.[27]

According to Japanese anthropologist Naomichi Ishige, the traditional method of making *takuan* is as follows:

> Dried daikon are packed tightly together in a wooden cask, and each layer is sprinkled with a mixture of rice bran and salt before being covered with another layer. The cask is then covered with an inner lid pressed down with a heavy weight.[28]

Depending on the size of the *daikon,* it should stay submerged in the rice bran mixture for a minimum of a month or longer.

Although the majority of commercially produced *takuan* sold outside of Japan is often dyed an artificially bright yellow with the addition of sugar, artificial sweeteners, and monosodium glutamate, the traditional bright yellow color comes from the seed pods of a type of gardenia called *kuchinashi no mi*.[29] Although traditional, un-dyed *takuan* is light brown and very wrinkly, it is by far more complex and concentrated in flavor.[30]

Finally, *kiriboshi* daikon is the completely dried strips of daikon resembling brown strands of twine or raffia. It must be fully reconstituted before use and it is usually braised with other vegetables or meats until tender to the bite. The flavor of the *kiriboshi* daikon is full of umami and is somewhat nutty with a very firm, crunchy texture. Its convenience as a dried item gives it a long shelf life and can be reconstituted as needed during any season to enhance or enrich a meal.[31]

Symbolism and anthropomorphism

Held annually, the Kanamara Festival, also known as the 'Festival of the Steel Phallus' in Kawasaki (about an hour from Tokyo), celebrates fertility and dates back over 300 years during a time when prostitutes asked for protection from a threatening syphilis epidemic.[32] It is not surprising to discover that the daikon is included in the festivities owing to its natural phallic shape, which has been written about in folk and religious stories for centuries.[33] In the festival, large daikon are carved by members of the community into phallic symbols and at the end of the festival they are auctioned off to whoever is willing to pay to bring them home; the money from the auction is donated to HIV/AIDS causes.[34]

Dokonjo Daikon: The Radish with the Fighting Spirit

Members of the Tokyo University of Agriculture (otherwise known as Tokyo Nōdai) perform their daikon dance called *Aoyama Hotori* on opening day as well as at other times throughout the year. A group of young men in black military-like uniforms chant and perform traditional Japanese cheering while holding either one or two very large daikon while thrusting their arms and legs and waving the daikon in school pride.[35] The daikon is their 'mascot' as well as a strong connection to agriculture. There are a few universities that use vegetables as mascots such as Delta State University with their 'Fighting Okra' or North Carolina School of the Arts' 'Fighting Pickle' but it is probably fair to say that vegetable mascots are unusual.[36] What distinguishes Tokyo Nōdai is the remarkable fact that they dance with the daikon and its inclusion is part of their university's identity.[37]

Dokonjo Daikon was not the first time that daikon were anthropomorphized in Japan. Takara Toy Company and Kiddy Land created a phenomenon (pre-dating the Daichan 'incident' by a few months) with *Aokubi* Daikon, a plush toy character.[38] This line of cute and cuddly daikon comes in many sizes with arms and legs, clothing accessories, and sometimes even holding and drinking a cup of green tea or plastic versions sitting in a hot bath! There is every imaginable trinket for sale with cute daikon on it – stickers, mobile phone charms, chopsticks, purses, pens and more. Although Japan is the land of cute and kitsch, the usual objects that are anthropomorphized are animals and foodstuffs like breads or burgers. *Aokubi* Daikon has definitely made its way into Japanese pop culture. One final example is the existence of a Flickr™ website group dedicated to photos of stuffed daikon toys in various anthropomorphic poses.[39]

Conclusion

The Japanese have a passion for vegetables that may defy belief by other cultures. The fact that the daikon is a significant ingredient of the Japanese diet and is seen as a popular icon most likely has a strong influence on their fondness for it. Perhaps their deep-rooted Buddhist background teaches appreciation of all life forms regardless of whether helping a radish in asphalt seemed ridiculous to many people, particularly overseas. In a time where technology is speeding up the pace of life, it is endearing to see that the people in Aioi could slow down and care for this daikon. Moreover, Daichan's struggle to survive conveyed a message that lifted the townspeople's spirit. Enshrining Daichan in children's books will continue to bring a smile to the Japanese people even after generations of Daichan's descendants have gone into vegetable obscurity. Besides, in children's books, it is the fundamental message of not giving up and having strength that matters, not whether the protagonist is a vegetable or not. The fact that the daikon is a very important vegetable in Japan enables children to more readily identify with the message that the book is passing on.

Dokonjo Daikon: The Radish with the Fighting Spirit

Notes

1. Richard Lloyd, The Times Online, 'Crying shame as streetwise giant radish is cut down in its prime,' <http://www.timesonline.co.uk/tol/news/world/article591348.ece>, [accessed 18 March 2008].
2. Jonathan Head, BBC News Online, 'Japan tries to save giant radish,' <http://news.bbc.co.uk/2/hi/asia-pacific/4677262.stm>, [accessed 18 March 2008].
3. Ibid.
4. 相生市, 兵庫県, ダイチャン, <http://www.city.aioi.hyogo.jp/new/daichan.pdf>, [accessed 1 July 2008].
5. Ibid.
6. Ibid.
7. 相生市,, 兵庫県, よっちゃんのここだけのはなし, <http://www.city.aioi.hyogo.jp/yocchan/064/index.html>, [accessed 1 July 2008].
8. Rubatzky and Yamaguchi, 409.
9. Rubatzky and Yamaguchi, 410.
10. Food Culture in Japan, 82–83.
11. Slow Food Foundation for Biodiversity, Ark of Taste, *Hanazukuri Daikon*, <http://www.slowfoodfoundation.net/eng/arca/dettaglio.lasso?cod=697&prs=0>, [accessed 8 July 2008]. Japan is not immune to the worldwide loss of biodiversity in foods and as more farmers are choosing to grow this high yielding variety, there is concern from small-scale farmers who are passionate about preserving unique regional varieties.
12. Quickly brined daikon tops are delicious as well as nutritious and were a regular part of my meals as a child.
13. Andoh, 118, Andoh, 215.
14. Kamamura, 112.
15. Kamamura, 21.
16. Yoneda, 33.
17. Yoneda, 167–171.
18. Ishige, 253.
19. Peter and Joan Martin, 178–179,
20. Ishige, 253.
21. Richie, 83. Although the diner may have had a variety of delicacies throughout the meal, most Japanese will not feel fully sated unless rice and pickles are served somewhere in the meal which is why it is served at the end as a sort of 'finale'.
22. Richie, 85.
23. Michael Ashkenazi and Jeanne Jacob, 113.
24. Richie, 85.
25. Richie, 85.
26. Ibid.
27. Ishige, 254.
28. Ishige, 254.
29. Finding high-quality artisan-made *takuan* was not as common in the marketplace in America when I was a child so we were forced to eat cheaply manufactured western variations. Today, there are many places to buy good-quality *takuan* outside of Japan.
30. According to Naomichi Ishige, the pigment comes from the rice bran pickling medium. Some households add citrus peels, dried persimmons, bonito flakes or dried eggplant leaves for additional flavor.
31. *Kiriboshi daikon* is useful as a nutritious 'filler' in many dishes and helps housewives in Japan stretch their budget in the kitchen.
32. Yoko Kubata, Reuters, 'Phallic festival celebrates fertility in Japan,' <http://www.reuters.com/article/lifestyleMolt/idUST32547520080407?pageNumber=2&virtualBrandChannel=0>, [accessed 9 July 2008].

33. E. Dale Saunders, Reviewed work of 'History of Phallicism in Japan' by Nishioka Hideo in *The Journal of Asian Studies,* Vol. 18, No. 2, (Feb. 1959), 292.
34. Yoko Kubata, Reuters, 'Phallic festival celebrates fertility in Japan,' <http://www.reuters.com/article/lifestyleMolt/idUST32547520080407?pageNumber=1&virtualBrandChannel=0>, [accessed 9 July 2008].
35. Tokyo University of Agriculture, <http://www.nodai.ac.jp/english/index.html>, [accessed 1 July 2008], *Aoyama Hotori* means 'beside the green mountain' and unfortunately I could not understand the words to the chant because it was being belted out in archaic Japanese.
36. Delta State University, <http://www.deltastate.edu/pages/1073.asp?item=1827>, [accessed 8 July 2008], North Carolina School of Arts,<http://www.ncarts.edu/pressreleases/Releases2007/Aug07/pickleonmsn.htm>.
37. Tokyo University of Agriculture, <http://www.nodai.ac.jp/english/index.html>, [accessed 1 July 2008].
38. Takara Tomy Toy Company, <http://www.takaratomy.co.jp/company/english/>, [accessed 9 July 2008].
39. Flickr, 'Daikons of Japan!' Group, <http://www.flickr.com/groups/348073@N21/>, [19 March 2008].

Bibliography

Andoh, Elizabeth. *Washo.* Berkeley: Ten Speed Press, 2005.
Ashkenazi, Michael & Jeanne Jacob. *Food Culture in Japan.* Westport: Greenwood Press, 2003.
Cwiertka, Katarzyna J. *Modern Japanese Cuisine.* London: Reaktion Books Ltd., 2006.
Kawamura, Yutaka. *The Pure Heart.* Mountain View: Ishi Press International, 1990.
Martin, Peter & Joan, *Japanese Cooking.* Middlesex: Penguin Books, Ltd., 1970.
みやざき，あゆみ，がんばれ大ちゃん, (恒星出版, 2006/03).
Ishige, Naomichi. *The History and Culture of Japanese Food.* London: Kegan Paul Ltd., 2001.
Richie, Donald. *A Taste of Japan.* Tokyo: Kodansha International Ltd., 1985.
Rubatzky, Vincent E. & Mas Yamaguchi. *World Vegetables: Principles, Production and Nutritive Values.* New York: Chapman & Hall, 1997.
Yoneda, Soei. *Good Food from a Japanese Temple.* Tokyo: Kodansha International Lt. 18.

The Pomtajer

Karin Vaneker

When Columbus discovered South America, roots and tubers were an important part of the local diet. Apart from cassava (manioc) and sweet potatoes, Spanish chroniclers reported the cultivation of pomtajer (Latin *Xanthosoma* spp.). Although an important indigenous food crop, little attention has been paid to this tuberous root since the Columbian Exchange. Cultivation and consumption was limited to people in and from (sub-) tropical regions. In the Netherlands, pomtajer serves as a basic ingredient for 'pom,' a popular oven dish within the Dutch Surinamese community. While frequently mentioned in literature and cookbooks, recipes are hard to find. Overall, the standard advice is that pomtajer is interchangeable with any root and/or tuber. This paper discusses its history, the use, and significance, as well as the culinary possibilities of pomtajer.

Roots and tubers play a significant role in the global food system, providing energy and carbohydrates. According to the International Food Policy Research Institute (IFPRI 2000), over 2 billion people in developing countries produce and consume major roots and tubers such as cassava, potato, sweet potato, and yam because of their high nutritional value. Since the 1960s and '70s the demand for roots and tubers has increased. This trend is projected to continue until 2020.[1] Such is also the case with araceae or aroids – currently consumed by around 400 million people, particularly in the (sub-) tropics. Taro (*Colocasia esculenta*) and pomtajer are the most widely grown and consumed aroids. Taro originates in south-east Asia while pomtajer is the only indigenous American aroid that is widely used for food (Bown, 258, 261). In addition, the Food and Agriculture Organization of the United Nations (FAO) states that pomtajer is attaining world importance as an energy food, in part because the nutritional value of 'the New World analogue of *Colocasia*' is comparable to the potato.[2] The small size of the starch grains makes pomtajer highly hypoallergenic and therefore probably easier to digest. Furthermore the mineral content is higher than taro (Hernández Bermejo and Léon), and the pomtajer is a more robust and disease-resistant plant.

There are 30 to 40 different kinds of pomtajer, which grows in (sub-) tropical climates and needs an annual rainfall of at least 1400 mm. It can reach near 2 meters in height and tolerates a certain amount of drought and shade. Compared to taro, the yield is heavier on drier soils. The large stem (corm) is harvested after 9 to 12 months (Bown, 259–261).

When thoroughly cooked, not only the corms and cormels but also the stalks, leaves, and even the petioles are edible. Still, in most (sub-) tropical regions not all parts of the plant are used and/or seen fit for human consumption; it is the corms and cormels that are most frequently eaten.

The Pomtajer

Under-utilized
Although regarded worldwide as one of the six most important root and tuber crops, pomtajer is categorized as an under-utilized crop that, thus far, has received scant research or attention (Reyes Castro, 3). Cookbooks provide little information and/or recipes. When the tuberous root is mentioned, the standard advice is that roots and tubers of any kind are inter-exchangeable. Boiled, mashed, roasted, fried, or steamed, as flour, chips, pancakes, or fritters, at home, in gastronomy or the food industry, pomtajer can be used in much the same way as the potato.

Tuberfusion
Apart from the Latin nomenclature, the names for pomtajer tend to be very confusing. Eddoe, tajer, cocoyam, malanga, and taro: pomtajer and taro have several overlapping names. In the Caribbean they are known as eddoes, while West Africans refer to both as cocoyam. Each plant of the aroid family has its own origin and botanical distinction. Taro is the general accepted term for *Colocasia*, yet *The Cambridge World History of Food* (Kiple and others 2000) uses the generic Austronesian term 'taro' for the four different roots belonging to the family of aroids: 1. Colocasia Taro: true taro, *Colocasia esculenta* (*Colocasia antiquorum*); 2. Alocasia Taro: false or giant taro, *Alocasia macrorrhiza*; 3. Cryptosperma Taro: giant swamp taro, *Cryptosperma chamissonis*; 4. Xanthosoma Taro: *Xanthosoma sagittifolium* (*Xanthosoma spp.*). Adding to the ongoing tuberfusion, most people, especially in the Western Hemisphere, can hardly tell the difference between the many different wild and cultivated species of tuberous plants including aroids and sweet potatoes.

Pomtajer: the plant and its origins
Like most roots and tuberous plants, this aroid has left little visible archaeological evidence, which makes it hard to trace its exact origins. At the time of the Colombian Exchange, pomtajer was cultivated from southern Mexico to Bolivia and in the greater Caribbean basin. Contrary to popular belief, the bulk of the diet of Amazonian-influenced cultures did not consist of meats but rather of starchy crops (Castaneda Langlois, 13). After cassava and sweet potato, pomtajer belonged to the basic planting of starchy root crops in the *conuco*, the elaborate farming system of the Taínos (or Arawaks), the pre-Columbian inhabitants encountered by Columbus when he first set foot in the Caribbean (Peregrine and others, 223).[3] Compared to maize and the sweet potato, the cultivation of pomtajer required considerably more labour.[4] One of the first accounts about South American roots and tubers comes from Columbus's aristocratic shipmate, Michele de Cuneo, who writes about 'roots like turnips, very big and in many shapes, absolutely white of which they made bread' (Parry, 71).

In Europe, roots and tubers like turnips were part of the staple diet of the commoner. So when the Americas were discovered, the European elite considered these to be inferior foods. Also, the Spaniards were accustomed to eating grains as a staple of their diet. Thus, though an important part of the local diet, the explorers never considered roots and tubers to be interesting enough to pay much attention to. This can also be inferred from De Cuneo's[5] description that the Caribbean 'turnips' are absolutely white (Parry 1997, 70, 71). Through the eyes of Europeans, probably all 'turnips' looked alike. Furthermore, De Cuneo writes that turnips were eaten both raw and cooked. Both pomtajer and cassava contain bitter juices. Only some varieties of sweet cassava contain a small amount of acid and can be eaten raw; like most roots they are usually cooked before consumption. When eaten raw, bitter cassava is highly poisonous. De Cuneo also might have described jícama (*Pachyrhizus tuberosus*) or ajipa (*Pachyrhizus ahipa*), which are typically eaten raw. De Cuneo also observes that one root (most probably cassava) is the main food for the islanders (Parry, 70,71 and Dunmire, 86). Even though it remains unclear which roots and tubers De Cuneo describes, his eyewitness account clearly indicates that the local diet consisted of more than one kind, and because he compares these with bread he recognizes them as staple foods.[6]

Apart from the sweet potato none of the other roots and tubers were appetizing or interesting enough to merit closer attention. Further, in the centuries that followed colonists had little interest in local foods and often claimed that they could not eat 'Indian food.'[7] This was still the case at the beginning of the eighteenth century, when the British woman, Gertrude Carmichael, accompanied her husband to the West Indies. Carmichael observed that *eddoes* were the only vegetables Africans used when preparing meat with peppers.[8] About these so-called *eddoes* she writes: 'It abounds upon every estate. The root is not unlike a rough irregular potatoe: – the leaves make excellent wholesome greens, and the negroe with addition of a bit of salt, or pork, has an excellent pot of soup…' (Mackie, 73, 74). Whether Carmichael writes about the indigenous pomtajer or the south-east Asian taro, or both, cannot be inferred from the text. Made at the height of the slave trade, her observations indicate the importance of *eddoes* in the local diet and, at the same time, confirm these were not important in the cuisine of the colonial powers.

Tubermigration

Before the twentieth century not much is known about the history of aroids. In an article in *Natural History*, Raymond Sokolov describes how taro reached and established itself in West Africa long before 1492. As a result of the slave trade taro was naturalized in the islands of the Caribbean. Reflecting the general knowledge and assumptions that are made about taro's arrival in the New World, Sokolov also confirms how, from the very beginning, the imported taro and local pomtajer were exclusively cultivated by, and were a part of, the diet of both the West African slaves and local population. Taro arrived with the slaves and became a popular 'ethnic' food. And vice versa, pomtajer frequently served as a provision on ships that, between the sixteenth and seventeenth centuries, headed for the Slave Coast. From there it spread further into West Africa and in some areas eventually became more popular than taro.

The Pomtajer

Surinam

From 1667 until 1975 Surinam was a Dutch colony; today it is one of South America's smallest countries with a highly diverse population. This is reflected through its cuisine, which is a mixture of ingredients and cooking techniques from local Indians, colonial powers,[9] African slaves, and Asian immigrants. At the end of the eighteenth century, the majority of the Surinamese population consisted of around 50,000 West African slaves of which the majority was kept on approximately 500 plantations.[10]

As a plantation economy Surinam had a very strong focus on growing cash crops such as sugar, coffee, and cacao. As elsewhere in the greater Caribbean basin the diet of the slaves was very poor. According to the medical doctor F.A. Kuhn, even as late as 1928 the diet of the former slaves still consisted of hard-to-digest starchy fruits and roots along with rice. Before the abolition of slavery in 1863 and in order to augment their poor provisions, slaves were allowed to cultivate vegetables and fruit on so-called *kostgrondjes* – small plots of land or home gardens, which they worked in their limited free time. Although little is known about the cultivation of pomtajer,[11] in Surinam the indigenous tuberous root grew in abundance (Benjamins and others).[12] As a result of the introduction of cash crops such as coffee and cacao, after 1750 pomtajer became a popular shade crop on many plantations (Bakker 33).[13] Referred to as *tayer*, it is frequently mentioned in eighteenth-century inventory lists of plantations thus making it available to slaves and others.[14] Such is the case in Pepperpot, which is one of the oldest Surinamese plantations established by the British before Dutch rule. Coincidentally pepperpot (or pepper pot) is a well-known savoury stew and/or soup in British Guyana and the greater Caribbean basin. Various sources identify the still-popular dish as of Amerindian origin (Mackie, 34 and Davidson et al., 596). In the region there are numerous pepperpot variations.[15] The ingredients and thickness of the dish varies but also today most recipes make use of hot peppers and cassareep: a thick boiled juice (syrup) of poisonous cassava that Amerindians either used as a preservative or as a cooking liquid. Depending upon the season and availability, various meats and vegetables were added and boiled in the pepperpot and, adding more meats and vegetables, it was frequently reheated. It is more than likely that pomtajer served as an ingredient. In addition, the techniques by which Surinam's indigenous population prepared roots and tubers were restricted to boiling and/or grilling.[16]

Although since 1916 around 20 Surinamese cookbooks have been published, the only Surinamese cookbook containing a recipe for pepperpot is *Groot Surinaams Kookboek* (1976, 148).[17] This book is considered a reference work but hardly provides information about the origin of the ingredients, recipes, and preparation methods employed by Surinam's indigenous population. Like most cookbooks from the Caribbean,[18] its focus is not on authentic dishes but on copying the domestic cuisine of colonial powers. Although *Groot Surinaams Kookboek* does not list it as an ingredient for pepperpot, the book contains a recipe for pom. Today one of the best-known and most popular Surinamese dishes[19] and also the only dish[20] that uses the corm of pomtajer.[21]

The Pomtajer

Culinary heritage

Within the community it is considered a festive dish and frequently referred to as a dish of Creole and/or Jewish origin. Ingredients such as chicken and citrus juice, its preparation technique, and the use of the oven make it very unlikely that pom is either of African or Amerindian origin because these major ingredients and the oven were introduced by the colonial powers. Until the beginning of the twentieth century an oven was neither common nor frequently used in Surinamese cooking. The preparation of oven dishes was restricted to professionals and/or wealthy households in which slaves were assigned to work in the kitchens. Over a quarter of the plantation owners were Jewish and, following a range of dietary laws, using an oven to prepare dishes. Pomtajer is very much suited to slow cooking in one-pot dishes and/or the oven. Due to a trade boycott, potatoes[22] were substituted for pomtajer[23] and pom was created. After this fusion of various cuisine and ingredients from Amerindians, colonial powers and African slaves, every ethnic group that migrated to Surinam adopted pom and started preparing its own slightly different version.[24] As a mixture of cuisine's and cultures from different continents, pom is an example of the mélange dish nowadays characterized as authentic. Both the dish and its major ingredient pomtajer migrated to the Netherlands.

The silent migration of pomtajer

Before and after independence in 1975, many Surinamese arrived in the Netherlands, where currently 335,779[25] Surinamese are living (compared to 475,996 people in Surinam[26]). Like the explorers several centuries before them, upon arrival they started to replicate their traditional foods. In order to be able to prepare Surinamese cuisine, ingredients such as pomtajer were imported from their homeland. Pom can be prepared with either fresh or frozen pomtajer. Especially towards the end of the twentieth century, the commercial production and distribution of grated and frozen pomtajer facilitated its export from Surinam to the Dutch Surinamese community.[27] Over 30 years after their arrival, and similar to that other former Dutch colony, Indonesia,[28] Surinamese cuisine and pom have been adopted outside the Dutch Surinamese community. A derivation of the national dish is *broodje pom* (sandwich pom), which is gaining popularity outside the community. Nowadays most of the 120 plus Surinam eateries and lunchrooms in Amsterdam serve either pom or *broodje pom*. It can even be ordered in Dutch take-away shops and delivered to your home.

Exploring the possibilities of pomtajer

Apart from recipes for pom, which appear mainly in Dutch, modern English-language cookbooks hardly ever contain recipes for pomtajer. As noted, generally the advice in these books and in relevant literature is that roots and tubers are interchangeable. Therefore, as an experiment and part of the exhibition in ImagineIC[29] in Amsterdam (21 May–31 August 2007) Michelin-starred chefs, food artists, and a food designer were invited to create new dishes with pomtajer.[30] In order to inspire the participants, Mavis

The Pomtajer

Hofwijk,[31] a well-known Amsterdam-based Surinamese caterer, demonstrated how to cook pom. This resulted in a pom clafoutis, created by Thorvald de Winter of restaurant Apicius, the only Michelin-starred restaurant that serves pomtajer. By modernizing his wife's family recipe, one Michelin-starred chef, Soenil Bahadoer,[32] also put his gastronomic version of pom on the menu in his restaurant De Lindehof. Inspired by Dutch snack food culture, food designer Katja Gruijters presented *pom de friet* (French fries made with pomtajer) and the *pom de kroket* (croquette). As a reference to his own half-German and the multi-ethnic background of the Surinamese population, food performer/artist Fredie Beckmans served pomtajer pancakes with Chinese chicken feet. American-born artist Debra Solomon used her own and pom's Jewish origin to prepare *latkes* with pomtajer and wild herbs. According to Oumar Mbengue Atakosso, *boulettes* are a traditional preparation method in his home country Senegal and he therefore created *boulettes de pom*.[33] Furthermore, inspired by the exhibition and experiment, Pay-Uun Hiu of the Dutch newspaper *De Volkskrant* published a recipe for vegetarian pom curry. The success of the experiment confirmed the interchangeability of roots and tubers. But when compared to the potato, for which exist many recipes and preparation techniques, the potential of pomtajer both in domestic cuisine and gastronomy has yet to be explored.

The use of pomtajer in South American cuisines (a few examples)

Unlike the potato, all pomtajer parts – the stems and leaves, and even the petioles – are edible. In the Caribbean the leaves traditionally serve as an ingredient for *callaloo*, a popular soup and/or stew. In Puerto Rico, *alcapurrias*, fritters or pancakes of pomtajer and plantain, are very popular both at home and as street food. Traditionally consumed boiled in soup, mixed with other vegetables and pork, chicken, fish, or beef, in Nicaragua the cormels of pomtajer are also peeled, boiled and pounded to a pulp with butter, milk, spices, and salt (Reyes Castro 15). For Cubans, *La malanga ha emergido como reina de las viandas* (pomtajer has emerged as the queen of the stew vegetables).[34] As such it is a very common ingredient in the popular multi-ethnic stew *ajiaco* (Cushman). In the 1960s pomtajer migrated from Cuba to the United States where it became available year-round in Hispanic markets. It is grown in Florida on a commercial scale, where fields of pomtajer are now a common sight (Bown and Meadows).

Africa

Both as a food and a crop, pomtajer has travelled beyond North America and Europe. In the nineteenth and twentieth centuries it spread into Asia and the Pacific,[35] while its cultivation in West Africa continues to increase (Bown). In West African cuisine, roots and tubers like pomtajer are traditionally boiled whole in their skins. Other common methods included pounding, roasting, and baking. Especially in Cameroon, the many forms in which it is traditionally prepared, processed, and consumed by various ethnic groups reflect the importance of pomtajer as a food crop. Cormels are peeled, boiled,

The Pomtajer

and eaten with a vegetable sauce. Together with palm oil, crayfish, salt, and pepper, the grated peeled corm and young leaves, the dish *ekwam* is regarded as a delicacy. This is also the case with *akwacoco*,[36] a traditional staple for the Bakweri people from the Southwest Province of Cameroon.[37] *Belbach* is a thick sauce prepared from young tender unopened leaves and petioles. In *nyeh* (bell soup) young leaves are used as a vegetable. These are also an ingredient for *kohki* or *koki* (referred to as cake or bread), made from ground cowpeas and ground corn or pomtajer. *Kohki*-beans are eaten with boiled cormels or plantain. In the colonial period *Cocoyam Koki* or *Endeley Bread*[38] made with pomtajer was the school-snack of children. In Cameroon the corms or cormels are also peeled, boiled and pounded into *futu* (see footnote 31), and finally, the cormels are used to prepare porridge (Onokpise et al 1999, 394).

It is as a result of pom's silent migration into the Netherlands and subsequent exploration of its potential in domestic cuisine and gastronomy that such new recipes as pom clafoutis were invented. Five hundred years after Columbus discovered the Americas the fusion of cultures, cuisines and ingredients has not come to an end but is still in an ongoing and dynamic process.

Notes

1. Roots and Tubers for the 21st Century: Trends, Projections, and Policy Options Gregory J. Scott, Mark W. Rosegrant and Claudia Ringler May 2000. http://www.ifpri.org/2020/briefs/numbers66.htm (accessed June 26th 2007).
2. http://www.fao.org/docrep/t0646e/T0646E00.htm (accessed 18 June 2007).
3. A 'conuco' is a circular heaping of soil with a diameter of several meters. This agricultural system still is practised in the Caribbean.
4. The International Development Research Centre, Canada: Document 23. Ancient Clones. http://www.idrc.ca/en/ev-115021-201-1-DO_TOPIC.html (accessed 27 June 2008).
5. Michele de Cuneo accompanied Columbus on his 1493 voyage. In 1495 he wrote an eye-witness account which is regarded as the most comprehensive (Parry 1997, 69).
6. Parry 1997.
7. http://www.ucalgary.ca/applied_history/tutor/eurvoya/columbus.html (accessed 10 June 2006)
8. Singular: eddo and/or eddoe are both used for taro and pomtajer.
9. Due to its colonial past in Surinamese cuisine Spanish and Portuguese influence is neglectable.
10. 1791: 3,790 whites, 1,330 Jews and 1,760 free non-whites in Surinam (Stipriaan Van, 1997, 74–75)
11. Agricultural knowledge was shared by locals and slaves.
12. According to the 'Encyclopaedie van Nederlandsch West-Indië' (1914–1917) it is an easy to grow crop.
13. Also today (in Africa and Melanesia) pomtajer is still used a shade crop.
14. Surinamese know pomtajer as pomtajer, pongtajer, tayer and taya. Taro is known as tajer, dasheen and eddoe. Tayer is 87 times listed on eighteenth-century inventory lists of the National Archives of Surinam http://nationaalarchief.sr/zoek/query.nl.pl?query=tayer (accessed 20 June 2008).
15. For a recipe see: http://www.recipezaar.com/192669 (accessed 22 June 2008) several sites claim it as a typical dish of the Guyanas. On recipezaar.com it is recommended to serve the dish with roots and tubers such as cassava and eddoes. Jamaicans often add starchy roots and tubers to the soup and/or stew.
16. Apart from boiled and grilled, cassava was also grated and/or pounded. Grated cassavaroot is used for 'cassavebread', the extracted liquid for cassareep.

The Pomtajer

17. *De Beste Surinaamse recepten*, Fokkelien Dijkstra, 1993, contains a recipe for 'peperpot.' Most of the recipes in this book appear to be adaptations from existing cookbooks.
18. Creole cuisine book.
19. To indicate the popularity of pom, if a Surinamese is asked to name his or her favourite dish, nine out of ten times the answer will be pom. Furthermore 'without pom there is no birthday' is a well-known Surinamese saying, which also specifies an occasion when pom is eaten.
20. The *Encyclopaedie van Nederlandsch West-Indië* (1914–1917) also refers to pom as a pie and describes the dish as follows: 'the big tajer, of which the stalk grows above the earth, is grated and treated with the juice of bitter oranges, afterwards with chicken or fish, made into a pie, which dish is known as 'pom'.'
21. In Surinam, leaves and/or stems (stalks) are not used, only the corm (exclusively for pom) and sometimes the cormels (soup) are used. Recipes that call for tajerblad (tajerleaf) refer to taro.
22. In Surinam potatoes do not grow well, even today, are many times imported. Also in the past most Surinamese ethnicities (including the Dutch) were unfamiliar with using ovens, which traditionally are used by Surinamese Jews and professional bakers.
23. In the Netherlands pomtajer is sold frozen and grated. Fresh pomtajer is only available in winter. Around Christmas, many Surinamese consider it an obligation to use fresh pomtajer. Apart from this self-imposed obligation, it is time-consuming.
24. The basic preparation method for pom is: in a high, metal dish, put sautéed chicken pieces between two layers of raw grated pomtajer, which is mixed with citrus juice and a sauce made from oil and/or margarine, onions, celery, tomatoes, salt, pepper and nutmeg. Bake the dish in an oven for at least one hour or until the pom becomes golden brown.
25. Source: Dutch Bureau of Statistics CBS. http://statline.cbs.nl/StatWeb/publication/?DM=SLNL&PA=37325&D1=a&D2=0-4,136,151,214,231&D3=0&D4=0&D5=0&D6=l&HDR=G4,G2,G3,T&STB=G1,G5&VW=T (accessed 26 June 2008).
26. July 2008 est. Source: https://www.cia.gov/library/publications/the-world-factbook/geos/ns.html (accessed June 25th 2008).
27. Approximately 5 Surinamese companies export frozen and grated pomtajer into the Netherlands.
28. Indonesian cuisine already existed when the country became a Dutch colony. Surinam became independent in 1971 its cuisine is young and developing.
29. See: www.pomophetmenu.nl.
30. See: Thorvald de Winter (www.restaurantapicius.com) Soenil Bahadoer (http://www.restaurant-delindehof.nl/) Katja Gruijters (www.katjagruijters.nl) Debra Solomon (www.culiblog.org), Fredie Beckmans (www.worstclub.nl) and Oumar Mbengue Atakosso (www.oumarmbengue.com).
31. See: www.surinaamsbuffet.nl
32. Bahadoer is the only Michelin-starred chef of Surinamese (Hindustani) origin.
33. In Senegal pomtajer aka cocoyam is used for fufu, a West African porridge (or pudding) of starchy roots and tubers.
34. http://www.cubanet.org/CNews/y05/apr05/29a9.htm (accessed 26 June 2008).
35. In these areas pomtajer often replaces taro in traditional dishes.
36. For *ekwam* the ingredients are tied together with young pomtajer leaves. For *akwacoco* pieces of crayfish, palm oil, salt and pepper are boiled with grated peeled cormels (Onokpise et al 1999).
37. http://www.congocookbook.com/forum/viewtopic.php?t=933 (accessed 26 June 2008).
38. According to the Congo Cookbook forum (http://www.congocookbook.com/forum/viewtopic.php?t=933) and (http://www.bakweri.org/2004/02/auntie_kate_coc.html) especially the Bakweri people of Cameroon use pomtajer as a staple. (both sites accessed 26 June 2008).

Bibliography

Bakker E. ed., *Geschiedenis van Suriname*. The Netherlands, Zutphen: Walburg Pers, 1998.

Benjamins, Herman Daniël and Snelleman, John F. eds. 1914–1917. *Encyclopaedie van Nederlandsch West-Indië*. The Netherlands, Den Haag/Leiden: Martinus Nijhoff/E. J. Brill, 1914–1917

Bown, Deni. *Aroids: Plants of the Arum Family* (second edition). Portland, OR: Timber Press, 2000.

Cushman, Gregory T. 'Cooking a Cuban Ajiaco: The Columbian Exchange in a Stewpot.' *World History Bulletin*. Fall 2006. Vol. XXII No. 2: 8–13.

Davidson, Alan, editor. *The Oxford Companion to Food*. Oxford: Oxford University Press, 1999.

Dunmire, William W. *Gardens of New Spain: How Mediterranean Plants and Foods Changed America*. Austin: University of Texas Press, 2004.

Facciola, Stephen. *Cornucopia II: A Source Book of Edible Plants*. Vista, CA: Kampong Publications, 1998.

Hernández Bermejo J.E. and León J., eds. 'Neglected Crops: 1492 from a Different Perspective.' Rome, Italy: FAO. Plant Production and Protection Series. No. 26 (1994): 253–258, 1994.

Kiple, Kenneth F. and Ornelas, Kriemhild Coneè, eds. *The Cambridge World History of Food. Volumes 1 & 2*. Cambridge: Cambridge University Press, 2000.

Kuhn, F.A. *Beschouwingen van den toestand der Surinaamsche plantagieslaven; Eene oeconomische-geneeskundige bijdrage tot verbetering deszelven*. [Considerations about the condition of the Surinamese plantation slave; an economic-medical contribution for their augmentation.] Amsterdam, 1928

Mackie, Christine. *Life and Food in the Caribbean*. London,: Weidenfeld & Nicolson 1991.

Mario González, Oscar. 'La vianda está perdida.' [The stew vegetable is lost.] Cubanet *Prensa Independiente* 29 Abril, 2005. (http://www.cubanet.org/CNews/y05/apr05/29a9.htm)

Meadows, Jean. 'Florida Food Fare.' *Sarasota Herald-Tribune*. 19 July 2000.

Onokpise, O.U., J.G. Wutoh, X. Ndzana, J.T. Tambong, M.M. Meboka, A.E. Sama, L. Nyochembeng, A. Aguegia, S. Nzietchueng, J.G. Wilson, and M. Burns. 1999. 'Evaluation of macabo cocoyam germplasm in Cameroon', in J. Janick (ed.), *Perspectives on new crops and new uses*. ASHS Press, Alexandria, VA. http://www.hort.purdue.edu/newcrop/proceedings1999/pdf/v4-394.pdf Reprinted from 1999 (accessed 10 June 2008)

Parry, J. H. *The Discovery of South America*. London: Paul Elek , 1979.

Peregrine, Peter N. & Melvin M. Ember, eds. *Encyclopedia of Prehistory. Vol. 7: South America*. New York and Boston: Kluwer Academic/Plenum Publishers, 2000.

Reyes Castro, Guillermo. *Studies on cocoyam (Xanthosoma spp.) in Nicaragua, with an emphasis on Dasheen Mosaic Virus*. Doctoral thesis. Uppsala: Swedish University of Agricultural Sciences, 2006.

Rozières De, Babette. *De Creoolse Keuken: de wereldkeuken uit de Caraïben brengt het beste uit de Aziatische, Afrikaanse, Indiase en Europese culinaire tradities samen*. [Creole: recipes from the culinary heritage of the Carribean, blending Asian, African, Indian and European traditions.] The Netherlands, Houten: Van Dishoeck/Unieboek bv, 2007.

Scott, Gregory J., Mark W. Rosegrant, and Claudia Ringler. 'Roots and Tubers for the 21st Century: Trends, Projections, and Policy Options.' 2020 Brief No. 66. Washington D.C.: International Food Policy Research Institute (IFPRI). Peru, Lima: Centro Internacional de la Papa. May 2000.

Sokolov, Raymond. 'Monserrat's Secret Gardens.' *Natural History* April 1992: 72–75.

Starke, A.A. & Samsin-Hewitt M. *Groot Surinaams Kookboek*. Rotterdam: Stichting Kankantrie, 1976.

Stipriaan Van, Alex. 'An Unusual Parallel: Jews and Africans in Suriname in the 18th and 19th Centuries.' *Journal for Jewish Literature and History in the Netherlands and Related Subjects. Studia Rosenthaliana*. Volume 31-Number 1/2-1997: 74–93.

Vaneker, Karin. 'Cooking Pom.' *Petits Propos Culinaires* 83, 2007: 31–48.

Vaneker, Karin. 2007. *Pom op het Menu*. Utrecht: Gopher BV, 2007. Endnotes: 998.

A Vegetable Zodiac from Late Antique Alexandria

Susan Weingarten

In his *Christian Topography*, Cosmas Indicopleustes has left us his description of his Christian world.[1] Writing in Greek in sixth-century Alexandria, he objects to the concept of the universe as spherical. For him, the universe is shaped like the Tabernacle of the Children of Israel in the wilderness, a rectangular building roofed by a rounded vault. He quotes the verse from Isaiah 40.22: *Who hath made the heaven as a vaulted chamber and stretched it out as a tent to dwell in.* And he provides us with a drawing to make his point quite clear.

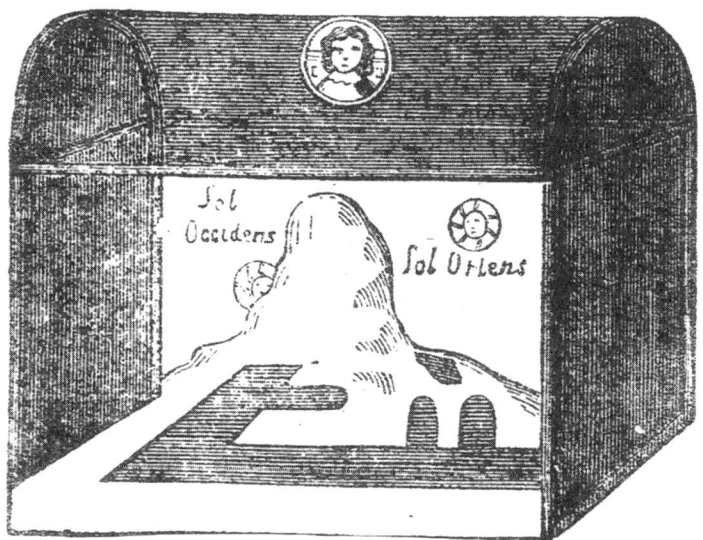

Figure 1. Cosmas' world-picture (from *Patrologia Graeca* 88,463).[2]

Inside the biblical Tabernacle was the table of the shewbread, with twelve loaves on it. This appears in Cosmas' text and as drawings, where the twelve loaves are identified with the twelve months of the year, and each corner of the square table with a season, just as another Alexandrian, Philo Judaeus, had done before him.[3] Within the universe, Cosmas' earth takes the form of a mountain, which rises within the Tabernacle, and the 'upper waters' are separated from the 'lower waters' as in the account in Genesis. Cosmas seems to have been unable to detach himself entirely from the old concept of the round universe, and his earth is encircled by a ring of the zodiac. Cosmas, indeed, has no less than three zodiacs illustrated in this book. The familiar 'pagan zodiac,' as he calls it, comes first, with its signs paralleled by the Alexandrian months, and illustrated with the usual figures: the ram, the bull, the heavenly twins, etc.[4] But Cosmas rejects this zodiac as pagan, and provides us with a second zodiac, with angels instead of the familiar signs. His audience was presumably reminded of the book of Revelation, where there is a vision of the heavenly Jerusalem descending to earth with 12 angels at its 12 gates, although Cosmas does not actually mention this. The angels are all very similar and are unnamed: perhaps Cosmas was uncomfortable at providing them with differentiating signs like pagan gods. We have no way of knowing, but it must have been inconvenient to have such similar signs for each month. Perhaps because of this problem, the *Christian Topography* provides us with a third image of a zodiac, called the Christian zodiac.

This zodiac is described as the 'crown of the year,' a quotation from Psalm 65.12 (LXX: Ps 64). This psalm describes Man coming to God's house and holy shrine. It praises God as creator and supplier of the universe: the ends of the earth and the distant seas (*loc. cit.* 6), the mountains, the peoples (7–8). There is also the cycle of the day: sunrise and sunset (9) and God's supplies of rain and harvest throughout the cycle of the year, which is 'crowned' by his bounty with blessed grain and herds (10–14).

Cosmas' Christian zodiac, his 'crown of the year,' shows different vegetable produce for each month. So instead of the name of the zodiac sign next to the name of the month, we have the name of a fruit or vegetable. I was born under the sign of Aries, the Ram, in mid-April, and I was delighted to see that in Cosmas' zodiac my vegetable sign is garlic!

There are many zodiacs in the ancient world, but I have found no others that consistently replace the familiar pagan signs with vegetables and fruits.[5] It is true, however, that in both Greek and Roman comedy there are a number of hints of associations of foods with the familiar signs of the zodiac: Athenaeus ii, 60 cites a comedy by Alexis where a huge hemispherical bowl representing the sky has foods arranged on it corresponding to the signs of the zodiac: 'fish, goats, with a scorpion between them, were laid out on the platter instead of the usual cheese, olives and hors d'oeuvres,' Alexis tells us. The bowl included slices of egg for the stars. This being comedy, I do not think we need to consider the question of whether people really ate scorpions. From Rome we have Petronius' satirical account of Trimalchio's feast, which included a circular plate with food corresponding to the signs of the zodiac, with double, often scurrilous interpretations of the familiar signs: the chef presents Aries the ram covered with chickpeas (*cicer arietes*) shaped like a ram's head, and later Aries is satirically re-interpreted by Trimalchio the host as representing horned professors and muttonheads. The chef interprets Gemini the Twins with pairs of testicles and kidneys; Trimalchio re-interprets them as teams of horses and oxen, lechers who are led by their balls and two-faced politicians.[6] These are both clearly satirical zodiacs. What serious classical author would associate these heavenly signs with something as mundane as food? But for a Christian author the simplest objects, if connected with God's plan, were worthy of the highest degree of attention.

These two Graeco-Roman zodiacs, then, appear to use actual foods as symbolic or interpretative in some way of the familiar zodiac sign. The same seems to be true of the use made of vegetable foods in various ancient calendar-type depictions of the Labours of the Months or personifications of months: for example, there is a mosaic in a sixth-century monastery in the Holy Land showing the Labours of the Months,[7] and a famous calendar from the fourth century illustrated by personifications of every month.[8] These figures, and others like them, occasionally have vegetable accompaniments, which act as typical and symbolic of the personifications: a bunch of grapes symbolizing the vintage or a wheat-sheaf for harvest-time. Grapes and wheat also typically accompany personifications of the four seasons.

Let us now turn to look a little more closely at Cosmas' twelve vegetable signs and think what they might signify for him and his contemporary audience.

Figure 2.

The vegetables and fruits of Cosmas' zodiac are seen in the pictures found in two out of the three extant manuscripts of the *Christian Topography*, now in the Laurentine

Library in Florence and the monastery of Saint Catherine in the Sinai peninsula, and are both dated to the eleventh century.⁹ There is a painting of the produce for each month, labelled in Greek with the name of the Egyptian month, its number, and the name of the vegetables depicted. The seasons are also named. The year, as in the Bible, starts in the spring.¹⁰

Thus we have:

1. Aries
(Pharmouthi) = March-April –>garlic *skoroda*

2. Taurus
(Pachon) = April-May –>wild barley *kinnai*

3. Gemini
(Pauni) = May-June –>Nuts and apricots *karya armenia*

4. Cancer
(Epiphi) = June-July –>wheat *sitos* and sycamore fig *skopymora* = perhaps *sykomora*

5. Leo
(Mesore) = July-August –>figs *syka* grapes *staphylia*

6. Virgo
(Thoth) = August-September –>olives *elaiai* and peaches *rhodakina*

7. Libra
(Phaophi) = September-October –>dates *phoinikes*

8. Scorpio
(Hathyr) = October-November –>asparagus *asparagoi*

9. Sagittarius
(Choiak) = November-December –>mallows *malachai*

10. Capricorn
(Tybi) = December-January –>endives *entybia*

11. Aquarius
(Mecheir) = January-February —>palm shoots *alatia*

12. Pisces
(Phamenoth) = February-March –>citrons *kitra*

Cosmas' text tells us that the zodiac illustrates the fruits of each month, for which we owe thanks to God (ix, 26). He picks out wheat, wine, and olive oil. These are, of course, the main ingredients of the Mediterranean food triad, which have been the

A Vegetable Zodiac from Late Antique Alexandria

staple foods of the area for generations. As such they are mentioned many times in the Bible (e.g. Deuteronomy 7.13, Psalm 4,8 [LXX], and *passim*). Cosmas points out that by God's grace these ripen in three successive months of the year (ix 27.10–14). Thus the signs of Cancer, Leo, and Virgo, from July to September, are changed to wheat, grapes, and olives. But these are only the basic foodstuffs, and as such may not express the immense bounty of the harvest of these months. In the pictured zodiac another food is added to each of them: sycamore figs[11] with the wheat, ordinary figs with the grapes, and peaches[12] with the olives. These all indeed ripen at these times, which confirms the vegetable zodiac as an accurate record of seasonal foods in Egypt.

Garlic, substituted for Aries in March-April, the first month of this zodiac, is well documented as an Egyptian crop from at least the second millennium BCE. Well-preserved garlic was found in Tutankhamen's tomb, while in the Bible, the Children of Israel in the wilderness, looking back nostalgically on the good and tasty food they had in Egypt before they were condemned to the monotony of manna, long for garlic (Numbers 11.8). Garlic also appears in Graeco-Roman Egyptian accounts as an important crop.[13] The late antique Jewish Talmudic sources are full of it: its potency, smell, flavour. I have added some details in an appendix. Although not directly relevant to our Egyptian zodiac, the Jewish sources give a very full picture of contemporary uses and perceived meanings in neighbouring countries. Today fresh garlic comes into Mediterranean markets in April. My local market sells it in five-kilo bundles.[14]

Other Egyptian crops have been discussed by Roger Bagnall in *Egypt in Late Antiquity*, using archaeological and papyrological evidence.[15] Egypt, subject to the annual flooding of the Nile, has three seasons: inundation, germination, and harvest. The papyri underline the importance of cereals (23–4). One common crop was barley, the food of the poor in common bread,[16] and of animals. The zodiac substitutes it for Taurus in April-May, the second month.[17] Wheat, the main crop, was harvested later, and we have already seen it in the zodiac instead of Cancer in June-July. Wheat was used for white[18] bread, which appears mostly in accounts of the more well-off households (24). After the wheat harvest came the inundation of the Nile. No further work could be done in the low-lying fields, but the time was ripe to harvest the tree crops: we have already seen the grapes, figs, olives, and peaches in the zodiac from July to September, and there is a date-palm instead of Libra in September-October. By the end of October the floods would have receded, and the vegetables of October-February in our zodiac (Scorpio, Sagittarius, Capricorn, and Aquarius) are all green shoots and leaves: asparagus, mallows, endives, and palm shoots or hearts. In some of the pictures they are shown bundled up, just as they were sold in the market. The winter-sprouting mallows are particularly interesting. Their Greek name is given as *malochia*, which nowadays refers to the mallow family of plants.[19] Here in Israel, mallows spring up after the winter rains from February, and are already dry and inedible by April. But a later, unseasonal rainfall may cause them to grow again. This happened during the siege of Jerusalem in 1948, when the inhabitants were suffering shortages of food, especially greens, and a late rainfall gave them supplies of mallows to eat as salads and cook in omelettes. In Cosmas' zodiac we see the same effect of late sprouting mallows caused by the inundation of the Nile. However, given that the zodiac comes from Egypt, *malochia* might refer to the plant known nowadays as *moloukhia*, *Corchorus olitorius*, still a valued part of the Egyptian diet, used to make a glutinous soup.[20] Endives, *Cichoria endivia*, were included in the Greek *lakhana* greens and/or *pikrides*, bitter herbs. Finally we have the produce of February-March instead of Pisces: the citron, depicted as large and knobbly, like the variety of citron still preferred today by Yemenite Jews for the Jewish festival of Tabernacles. Between the barley and the wheat, instead of the sign of Gemini, there is an unclear heading: *karya armenia*. Wolska-Conus suggested that this might refer to an otherwise unknown type of nut: Armenian nuts (on the model of *karya persika* = Persian nuts = walnuts). *Karya* are certainly nuts, but *armenia* can only refer to *armenika*, apricots, if we look at the colour reproduction of the zodiac which was not available to Wolska-Conus.[21] Apricots arrived in the Mediterranean basin in Roman times, unlike peaches, which had been around since the middle of the first millennium BC, and they would be enjoying their very short season at this time in many Mediterranean lands.

Bagnall warns us of the foodstuffs in the papyri that 'we would be ill-advised to swallow the menu whole.'[22] This documentation is biased towards the taxable crops and the diet of the richest stratum of the population. There seems to be a similar sort of bias in the zodiac vegetables. We know from other sources that legumes – chickpeas, beans, fenugreek and even lupines[23] – made up a major part of the food of the poor, contributing protein to a diet low or lacking in meat, but here they do not appear at all. Indeed the papyri in general, and the chance finds of excavations too, do not give a very clear picture of what people actually had to eat day by day in late antique Egypt.

However, there is one very specific papyrus archive that goes some way to filling this gap. This belonged to a tax official called Theophanes, who travelled from Egypt to Syria via Palestine in the fourth century, and left all the accounts of his daily purchases of food (with the Egyptian dates on which it was bought) in an archive found in Hermopolis in Egypt, and now in the John Rylands library in Manchester, and hence called Papyrus Rylands.[24] Theophanes began his journey in Hermopolis on the Nile, probably in March and seems to have sailed 200 kilometers downriver to Babylon at the apex of the Nile delta. From here he journeyed by land to Pelusium, and then followed the coast through Palestine and Syria to his goal: Antioch, one of the major centres of the late Roman Empire, where he stayed for two-and-a-half months. As an official traveller, he had his accommodation and horses paid for, but had to buy food for himself and his party, hence the accounts. I have recorded all the foods bought by Theophanes at the beginning and end of his journey when he was in Egypt, so as to see how they parallel the monthly foods of Cosmas' zodiac. We have many more details about purchases during his stay in Antioch, but I have noted only those that are relevant to the zodiac.

A Vegetable Zodiac from Late Antique Alexandria

Theophanes and Cosmas

Theophanes' purchases	Cosmas' vegetables in zodiac
PHAMENOTH (FEBRUARY-MARCH)	
?Hermopolis	citron
18 wine	
On the boat	
21 wine	
Babylon (in Nile Delta)	
27 common [=*kibarios*] bread, wine	
28, 29, 30 wine	
PHARMOUTHI (MARCH-APRIL)	
Babylon	garlic
1 wine	
2 wine, *lachana* [greens]	
3 artichokes	
5 *lachana* [greens], lettuces	
8 *lachana* [greens], *kemia*[25]	
9 *kemia*, artichokes	
Athribis	
11 *kemia*	
PACHON (APRIL-MAY)	
Antioch	[wild] barley
PAUNI (MAY-JUNE)	
Antioch	nuts and apricots?
10, 15, 19, 20 apricots	
22 apricots, dates	
22 garlic	
EPIPHI [EPEIPH] (JUNE-JULY)	
Antioch	wheat and sycamore figs
16 apricots	
25 apricots, grapes, ?more grapes	

A Vegetable Zodiac from Late Antique Alexandria

Theophanes' purchases   **Cosmas' vegetables in zodiac**

MESORE (JULY-AUGUST)

Raphia   figs and grapes
9 fine [=*katharos*] bread, common bread,
    wine, gourds, melon, sycamore fig, grapes,
    vermouth
Rhinocolura
10 fine bread, common bread, ordinary oil,
    wine, gourds, cucumbers, peaches
Ostrakine
11 cucumbers, grapes
Kasion
11 ordinary oil
Pentaschoinos
12 olives, sycamore figs, leeks
loaves for the boys
Pelusium
12 leeks, sycamore figs, common bread
13 fine bread, common bread,
    ordinary oil, vinegar, nuts
Skenai
15 fresh figs, loaves, wine
Heliopolis
16 green veg.
Babylon
16 *lachana* [greens]
17 fine bread, common bread,
    wine, vermouth, grapes
18 common bread
19 common bread
20, 21 soft bread
Ampelon/Hermopolis
22 common bread
23 common bread

THOTH (AUGUST – SEPTEMBER)

                                                                  olives and peaches

A Vegetable Zodiac from Late Antique Alexandria

As we are dealing here with vegetable foods, I have not recorded purchases of meat, fish, cheese, or eggs. The main food here is clearly bread, as it was in most of the ancient Mediterranean world. We note that the wheat and barley of the zodiac is here made into bread of two qualities: fine white *katharos* wheat bread for the master, common *kibarios* bread (probably barley bread) 'for the slaves.' This was common practice, legislated against by the rabbis in ancient Palestine.[26] The zodiac cites garlic as the harvest of March-April, when indeed it is harvested today. Theophanes, however, does not buy it until later, in May-June. We cannot know whether he had brought his own supply, which ran out, or whether it was harvested later around Antioch than in Egypt and Palestine. We saw the zodiac harvest for May-June was apricots and nuts. Apricots are indeed in season in June, even today a very brief season – barely six weeks – and we see Theophanes making the most of it in Antioch: he buys apricots every few days all through the month of June. Apricots are mentioned by Pliny as having arrived in Rome by his time (first century AD), but they have not been identified at all in the Talmudic literature. Perhaps they were still a scarce delicacy in fourth-century Egypt, which is why Theophanes buys them so often, but by the sixth century when Cosmas writes, they could have become established in Egypt. However, I have not found records of any archaeological evidence for their cultivation there.

Theophanes is still in Antioch in June-July. Here he is still able to buy sycamore figs, presented in the zodiac as the produce of the previous month, as well as grapes, which appear in the zodiac for the following month. And, of course, he is still making the most of the end of the apricot season. His last purchase of apricots, however, is on the second of the following month, Mesore, (July-August) when he is already on his journey home. Then as now, this is the height of the season for soft fruits. Theophanes enjoys a large variety, especially the fresh figs and grapes specified in the zodiac. As it is the height of the grape season, he has grapes every other day until Babylon, where he embarks on the boat to sail home up the Nile. He also buys wine, like bread, of different qualities. Finally, he finds some peaches on the Egyptian border earlier than the zodiac, which only mentions them for the following months of August-September.

Yet another document we can use to compare with the vegetable zodiac is the 'dietary calendar' of Hierophilus the sophist, which has been variously dated between the fourth and seventh centuries. We shall see below that there are grounds for thinking that at least some of this calendar originates in Alexandria. Hierophilus recommends bodily care, and foods to eat, and to abstain from, month by month. The calendar is written to correspond with the medical theories of the time, which saw the body as consisting of four humours, each of which was dominant in a different season of the year. Thus, for example, June governs 'hot blood,' so people are told to take cold foods and avoid bitter and dry flavours, and similarly for the rest of the months. Since food was seasonal as well, these recommendations had also to take into account what foods were available at different times of the year. However, the calendar is extant in different textual versions, which sometimes give contradictory advice.[27] Thus the versions used by Jeanselme and Koder forbid garlic in June, while Dalby's version recommends it. (Since many people saw garlic as an aphrodisiac, we should perhaps reject Dalby's version here, for the calendar adds: No love-making).[28] However, we recollect June was the month when Theophanes bought his garlic in Antioch. Garlic appears in the zodiac, it will be recalled, in April – and it is recommended then in Dalby's and Koder's versions, although Jeanselme's version recommends making only very light use of it. In July there is more agreement: the figs and grapes of the zodiac, which we know Theophanes bought to eat, are recommended by all versions, as are the peaches of August and the fresh dates of September. In January and February Hierophilus mentions citron – February in the zodiac.

This leaves us with the green vegetables: the asparagus, mallows, and endives which the zodiac attributes to October to January. Hierophilus similarly recommends asparagus for October, mallows for November, possibly endives for December and wild asparagus for January. However, the tenth-century farming manual called the *Geoponika*, compiled in Constantinople, suggests sowing the seeds in these months, not harvesting them to eat, and these vegetables are usually at their best in the spring all around the Mediterranean. The only place where these greens grow from October to January is in Egypt, following the inundation of the Nile. Thus we can probably attribute an Egyptian origin to Hierophilus or the source he used for these recommendations – and indeed Alexandria was a major centre of Greek medical writings.

Thus we have seen how the vegetable zodiac of Cosmas is true to its origins in Graeco-Roman Egypt, reflecting the fruits and vegetables eaten in season by local people, typified for us by the pagan Theophanes in the fourth century. Further perspectives can be seen in the dietary calendar of the sophist Hierophilus, in the *Geoponika*, a farming manual complied for a Byzantine emperor, and in the neighbouring Jewish Talmudic sources. The particular Christian meaning Cosmas gives to his zodiac, however, like his picture of the world, is uniquely his own.

Appendix: Talmudic garlic

Talmudic sources, collated from the third to the seventh centuries,[29] almost without exception praise garlic. Whereas coarse bread [*qibar*], new beer, and other vegetables were seen as bad for you, leeks were an exception – and garlic was even better (Babylonian Talmud [=BT] Eruvin 56a). Sold in bundles (Mishnah Peah vi 10 and parallels) or plaits (Tosefta Demai iii 12) in the market, garlic came in cultivated or wild varieties (*shum*; *shumanit*) and there was also 'Baalbec' garlic (Mishnah Ma'aserot v 8 and parallels). It often appears in lists of tasty or sharp vegetables together with onions, leeks and mustard (e.g. Mishnah Uqtzin i 2). It was sometimes laid out on the flat roof where the dew would keep it fresh (Mishnah Makhshirin vi 2 etc). It could be eaten whole, but someone who started eating it from the bulbs rather than the leaves was considered greedy and a glutton (BT Beitzah 25b etc). Garlic was also pounded in a mortar for use (e.g. Mishnah Tevul Yom ii 3), especially on Fridays for the Sabbath (e.g. Tosefta Shabbat v 3). It was added to *miqpeh* (Mishnah Nedarim vi 10 and parallels), the bean or lentil stew which was an important food, and mixed with it (BT Shabbat 140a), although it is not clear whether it was cooked with it or added raw. Certainly the taste was unmistakable – the Mishnah rules that someone who took a vow not to eat *miqpeh* was considered to have included abstention from eating garlic in his vow, as this was the main taste of the dish (Jerusalem Talmud [=JT] Nedarim 40a). Eating garlic on Sabbath eve was supposedly one of the ten rules laid down by Ezra the scribe: garlic was considered an aphrodisiac (one of the five things which increased sperm JT Ma'aserot 48c): and a husband was obliged to have sex with his wife at the proper time (BT Bava Qama 82a). Many men were absent during the week, so the Sabbath eve was prescribed as the proper time for marital intercourse. The Jerusalem Talmud added that garlic increased love and decreased forbidden lusts (JT Megillah 75a), while the Babylonian Talmud said it increased love and decreased jealousy (BT *loc cit*). It was also supposed to be effective against parasites in the guts (*loc cit*), but it was not considered good for the baby if a breast-feeding mother ate garlic (Sifre Numbers 89). Garlic was considered to be particularly popular with snakes – people had to beware lest they come and eat from a dish of garlic and contaminate it with their venom (Midrash Genesis Rabbah 54 21). According to the Babylonian Talmud, eating pre-peeled garlic was also dangerous, like drinking water that had been left uncovered at night, or sleeping in a graveyard, since evil spirits might be around (BT Niddah 17a). Eating garlic was seen as something characteristic of Jews (and Samaritans) by the third century Palestinian Mishnah complied by Rabbi Judah haNasi. (Mishnah Nedarim iii 10). The Babylonian Talmud, however, notes the unpleasantness of the smell on more than one occasion. 'Someone who eats garlic and becomes smelly, is he going to do it again?' asks Rav Hisda rhetorically (BT Berakhot 51a). Also in the Babylonian Talmud, there is a story told about Rabbi Judah haNasi, who smelled garlic in his study house and asked the offender to leave. Rabbi Hiyya left, and then all the rabbis followed. The next day Rabbi Judah's son asked Rabbi Hiyya if it was he who had eaten garlic, but he denied it – he knew it was not acceptable in the study house, he said. The story is told in a context of rabbis who admit to offences they have not committed to avoid embarrassing others. It is clear here that garlic is unacceptable. But we cannot know whether Rabbi Judah haNasi really thought this, like the Roman upper classes with whom he is strongly identified,[30] and unlike the general Jewish lower classes to whom he is probably referring in Mishnah Nedarim above, where garlic eating is seen as typically Jewish. It may be simply that this is an attitude attributed to him by the editors of the Babylonian Talmud, where, as we saw, the smell of garlic is now taken as unacceptable, at least in the study house.[31]

A Vegetable Zodiac from Late Antique Alexandria

Notes

1. For the various titles given to this work, see *Cosmas Indicopleustes: Topographie Chrétienne* 3 vols (ed. Wanda Wolska-Conus, Paris, 1968–1973) 59–61. This is the edition I have used. Also: Wanda Wolska *La Topographie chrétienne de Cosmas Indicopleustes: théologie et science au VIe siècle* (Paris, 1962). 'Indicopleustes' means 'the traveller to India'. The work ends with details of Indian flora and fauna, perhaps from a larger geographical work.
2. Cf. Umberto Eco *Baudolino* (tr. W. Weaver, London, 2002) p. 257.
3. Philo *Special Laws* i 172. cf. *Questions on Exodus* ii, 76.
4. The pagan zodiac appears in book iv, 15, and it is also mentioned in the text in vii, 89; ix, 10.
5. H.G. Gundel *Zodiakos: Tierkreisbilder im Altertum* (Mainz-am-Rhein, 1992), for example, only mentions Cosmas' pagan and angel zodiacs.
6. See: K.F.C. Rose, J.P. Sullivan 'Trimalchio's zodiac dish (Petronius, *Sat*. 35. 1–5)' *Classical Quarterly* NS 18 (1968) 180–184.
7. Monastery of Kyria Maria, Beit Shean Israel: *New Encyclopedia of Archaeological Excavations in the Holy Land*, (Jerusalem, 1993) vol. 1, sv Beit Shean.
8. See M.R. Salzman *On Roman time: the codex-calendar of 354 and the rhythms of urban life in Late Antiquity* (Berkeley/Los Angeles, 1990).
9. The Sinai MS seems to have been written in Cappadocia, while the Laurentine MS seems to originate from Mt Athos. The zodiac has not been preserved in the ninth-century Vatican MS from Constantinople.
10. Month of spring: Exodus 12.2.
11. Sycamore figs, Gk *sukamora*. Suggested by the editor for the ?garbled terms *kopumora/skopumora*. This was a species of fig which was typical of Egypt, (see Amos 7.14), apparently less tasty than ordinary figs *Ficus carica*.
12. *Persica vulgaris* is called *melon persikon* in classical Greek. Latin *duracinum* is the origin of the name *rhodakina* here.
13. D.J. Crawford 'Garlic-growing and agricultural specialization in Graeco-Roman Egypt' *Chronique d'Egypte* 48/2 (1973) 350–63.
14. I usually find someone else to go halves with: another Ashkenazi woman who thinks 5 kg is too much even for a year's supply.
15. R. Bagnall *Egypt in Late Antiquity* (Princeton, 1993). The page references here are to this book.
16. Greek *kibarios*, Aramaic *qibar*.
17. The Greek word used here is unusual: *kinnai*– found in Dioscorides of a Cilician variety – rather than the more usual *krithe*.
18. Greek *katharos*, white; Hebrew *neqiya*, clean.
19. The *Flora Palaestina* gives at least six varieties.
20. Claudia Roden *New Book of Middle Eastern Food* (London, 1986) 161–3.
21. Several symposiasts proposed that perhaps *karya armenia* referred to edible apricot kernels. However, the colour reproduction of the zodiac which I subsequently received, courtesy of the Laurenziana library in Florence, shows large brown oval nuts and *smaller* round orange apricots.
22. Bagnall *op cit* 23.
23. Lupines are poisonous unless soaked and boiled by lengthy processes.
24. J. Matthews *The journey of Theophanes: travel, business and daily life in the Roman East* (New Haven/London, 2006); text: C. Roberts *Catalogue of the Greek and Latin papyri in the John Rylands library at the University of Manchester* vol iv (1952).
25. *kemia* (and *kemoraphanos*) appear several times in Theophanes' accounts. Matthews, comparing them with the *cymae* of Diocletian's price edict, suggests they may refer to 'the young shoots of some member of the cabbage family.' On the other hand, Andrew Dalby in private correspondence has suggested they may be celery, based on a gloss on *chamairaphanos* as *apion* [=Lat. *apium*, celery] in Oribasius *Eclogae medicamentorum* 51.1.

26. Sifra Behar 7: There is a tannaitic source: *For it is good for your people* (Deuteronomy 15.16) Your people in food and your people in drink. For you should not eat fine bread and [your slave] eat common bread [*qibar*]… : D. Sperber '*Qibar* bread' *Tarbiz* 36 (1967) 13–14, in Hebrew.
27. I have used the texts of the Thesaurus Graecae Linguae, from A. Delatte *Anecdota Atheniensa et alia* vol 2 (Paris 1939) and J.L. Ideler *Physici et medici Graeci minores vol 1* (Berlin 1841 repr Amsterdam 1963); the translations of A. Dalby *Flavours of Byzantium* (Totnes, 2003) 52–56; 161–169, who uses Delatte; E. Jeanselme 'Les calendriers de regime à l'usage des Byzantins et la tradition hippocratique' in *Mélanges offerts à M. Gustave Schlumberger* vol I (Paris, 1924) 217–233 who uses the text of Cod Paris 2314 and 396, and the translation of J.-F. Boissonade 'Traîté alimentaire du médecin Hiérophile' *Notices et extraits des manuscripts de la Bibliothèque du Roi* vol 2 part 2 (1827) pp. 178–273; J. Koder 'Stew and salted meat – opulent normality in the diet of every day?' in L. Brubaker, K. Linardou (eds) *Eat drink and be merry (Luke 12.19) – food and wine in Byzantium* (Aldershot, 2007) 59–72, who uses the text of R. Romano 'Il calendario diatetico di Ierofilo' *Atti dell Academia Pontaniana* ns 47 (Naples, 1999) 197–222, which summarises three versions.
28. Emily Gowers *The Loaded Table* (Oxford, 1993) chapter 5: garlic breath, as well as the Talmudic appendix below.
28. For a brief explanation of the Talmudic literature, see my paper 'Nuts for the children: the evidence of the Talmudic literature,' in R. Hosking (ed.) *Nurture: Proceedings of the Oxford Symposium on Food and Cookery 2003*, (Bristol, 2004).
29. See Gowers, n. 28 above.
30. For Jews and garlic in more modern times: Maria Diemling '"As the Jews like to eat garlic": Garlic in Christian-Jewish polemical discourse in early modern Germany', in L.J. Greenspan, R.A. Simkins G. Shapiro (eds) *Food and Judaism* (Omaha, 2005).